THE RAPE VICTIM

THE RAPE VICTIM

by

Elaine Hilberman, M.D.
Department of Psychiatry, School of Medicine
University of North Carolina

A project of the Committee on Women of the American
Psychiatric Association:

Elissa Benedek, M.D., *Chairperson*
Mary Ann Bartusis, M.D.
Ann L. Chappell, M.D.
Virginia Davidson, M.D.
Nancy Durant, M.D.
Elaine Hilberman, M.D.

Approved for Publication by the Council on National
Affairs and by the Trustees of the American Psychiatric
Association.

AMERICAN PSYCHIATRIC ASSOCIATION

Basic Books, Inc., Publishers / New York

Copyright © 1976 by American Psychiatric Association
Library of Congress Catalogue Card No. 76-5627
ISBN: 0-465-06830-8
Printed in the United States of America
76 77 78 79 80   10 9 8 7 6 5 4 3 2 1

CONTENTS

ACKNOWLEDGMENTS

This monograph was conceived as a major project of the American Psychiatric Association Committee on Women, chaired by Dr. Elissa Benedek, and would not have been possible without the committee's support, invaluable critical review, and encouragement.

Special acknowledgments are extended to Dr. Ann Chappell for her case reports; Drs. Carol Nadelson and Malkah Notman for both their case report and substantive contribution to the chapter on the role of the psychiatrist; Dr. Anne M. Seiden for a case report and contributions to the chapter on child victims; and Dr. Elissa Benedek for her substantive contribution regarding children in the courts.

The author also wishes to thank Ms. Carol Motz for her tolerance and patience while typing and retyping the manuscript.

Finally, special gratitude is extended to the author's nine year old son, who has been, perhaps, the most formidable critic of all. He speaks for himself in this unsolicited note:

> I hate my mother for writing even so much as the first chapter of this book because it took so much of her time. However in conclusion, I love my mother and nothing can change that.
>
> *Joshua*

Elaine Hilberman, M.D.
Chapel Hill, North Carolina
January, 1976

INTRODUCTION

The problems of victims of sexual assault who are courageous enough to identify themselves as such are notorious. The act of reporting a rape initiates a most complex process. The victim is confronted with the usual institutional patterns of the hospital and the criminal justice system, which are likely to be perceived as confusing and alien. It is emphasized that she presents herself to the authorities at a time of crisis. It is a crisis which differs from other crises in that one's usual support system is more likely to be disrupted. Families, friends, and lovers, because of their own perceptions about rape, may desert and further isolate her. Additionally, the crisis is never limited to one's person since the victim, by the act of reporting, becomes public property, and is at the mercy of the hospital, police, courts, media, and community opinion. Rape represents an act of violence and humiliation in which the victim experiences not only overwhelming fear for her very existence, but an equally overwhelming sense of powerlessness and helplessness which few other events in one's life can parallel. The victim's needs are for empathy and safety, and for a sense of control over what has happened to her and what will happen to her in her dealings with hospital and law enforcement. In the absence of sensitivity to these needs, the experience of reporting becomes another assault.

The profound impact of the rape stress is best understood in the context of rape as a crime against the person and not against the hymen. Victims of violent crimes frequently experience a life crisis which more often than not goes unrecognized. Bard and Ellison[1] describe the spectrum of stresses confronting the victim depending on the extent of the personal violation. Burglary, for example, is experienced as a violation of the self in that one's home and possessions are symbolic extensions of the self. Armed robbery intensifies the stress by the added dimension of an encounter between the victim and the criminal. The self-violation is thus compounded by a coercive deprivation of independence and autonomy, in which the victim surrenders his/her controls under the threat of violence. An actual physical assault in addition to the robbery further stresses the victim for whom the injury to the body (or envelope of the self) serves as concrete evidence of the forced surrender of autonomy. Rape, then, becomes the ultimate violation of the self, short of homicide, with invasion of one's inner and most private space, as well as the loss of autonomy and control. In this schema, it becomes irrelevant to differentiate vaginal from anal or oral violation; it is the self and not an orifice that has been invaded. Thus, for the virgin, the prostitute, the housewife and the lesbian, the core meaning of rape is the same.

The women's movement has had a major role in delineating and confronting the abundant rape mythology, as it relates to societal attitudes about the assailant, the victim and the crime. Behavioral scientists in recent years have identified a body of information about rape, as well as an unfolding series of reactions experienced by the victim after the rape. It is the aim of this monograph to summarize what is now known about the needs and experiences of the victim and her family, and to provide a framework in which the clinician can more knowledgeably provide assistance and support to the victim. If the monograph seems overinclusive, it is a reflection of the extent to which societal attitudes have permeated every institution with which the victim comes in contact. It is hoped that the monograph will reinforce the need for clinicians to inform themselves about local hospital policy, criminal justice procedure, rape statutes, and community attitudes and services because it is not possible to adequately treat victims without an understanding of the context in which rape occurs.

References

1. Bard, M., Ellison, K.: Crisis Intervention and Investigation of Forcible Rape. *The Police Chief* 41:5, 68-73, May 1974.

RAPE DEFINED

Rape has been defined as the act of taking anything by force.[1] Rape is legally defined[2,3] as carnal knowledge of a person by force and against that person's will. Two elements are necessary to constitute rape: 1) sexual intercourse and 2) commission of the act forcibly and without consent. The slightest penetration by the male organ constitutes carnal knowledge; neither complete penetration nor emission is required. Force may be defined as the use of actual physical force to overcome the victim's resistance, or the use of threats which result in victim acquiesence because of fear of death or grave bodily harm.

Rape experiences encompass a wide spectrum, ranging from surprise attack with threats of death or mutilation to an insistence on sexual intercourse in a social encounter where sexual contact was unexpected or not agreed upon. In the latter case, non-consent is often overlooked or misinterpreted by assuming that certain social situations imply a willingness for a sexual relationship.[4] Indeed, sexual violence is a frequent and unreported event in the usual dating situation on college campuses as reported by Kanin[5] and Schultz and De Savage.[6]

It is recognized that male rape and statutory rape occur and present added dilemmas. The overwhelming majority of victims, however, are female, and this monograph will address itself only to the female victim of forcible rape.

References

1. *Oxford English Dictionary*, 1971 Ed.
2. Evrard, J.: Rape: The Medical, Social and Legal Implications. *American Journal of Obstetrics and Gynecology* 111:2, 197-199, September 1971.

3. Report of the District of Columbia Task Force on Rape. Subcommittee of District of Columbia City Council, July 1973.

4. Notman, M. T., Nadelson, C. C.: The Rape Victim: Psychodynamic Considerations. *American Journal of Psychiatry* 133:4, 408-413, April 1976.

5. Kanin, E. J.: Selected Dyadic Aspects of Male Sex Aggression. Reprinted in *Rape Victimology*, ed. LeRoy G. Schultz, Charles C. Thomas, Springfield, Illinois, 1975. (*Journal of Sex Research*, 5:1 February 1969).

6. Schultz, L. G., De Savage, J.: Rape and Rape Attitudes on a College Campus. Reprinted in *Rape Victimology*, ed. LeRoy G. Schultz, Charles C. Thomas, Springfield, Ill., 1975.

THE RAPE VICTIM

CHAPTER

ONE

THE SOCIOCULTURAL CONTEXT OF RAPE

Rape crisis programming is usually initiated only after a victim has been extraordinarily abused by the authorities, both medical and criminal justice, and the following example serves to illustrate this:[1]

> In late 1973, a child was brought to the emergency room of a major teaching and referral hospital by a distraught mother who gave the history that the youngster had been raped. The hospital, which had only vague procedural guidelines for the treatment of rape victims, informed the mother that her daughter would not be examined unless she had a warrant for the assailant's arrest. The parent was driven some twenty miles to the Sheriff's office where she was told that a warrant could not be issued unless the child was first examined, and medical evidence of rape confirmed. Back at the emergency room, a physician reluctantly examined the child, but refused to tell the mother the results of the examination. The appropriate and long overdue sequel to this chain of events was a public outcry which resulted in a mobilization of hospital, law enforcement and community resources to provide more effective services to rape victims.

How is one to find a framework in which to understand how a group of ordinarily well-meaning and empathetic individuals representing both hospital and law enforcement could have unwittingly collaborated in such inappropriate behavior? A review of the medical literature on rape through 1973 proved quite revealing. Most striking was the absence of any significant literature about the victim, other than that relevant to strictly medicolegal concerns, although there were a number of articles dealing with the need to understand and rehabilitate the rapist. The reader was provided with a mass of instruction designed for medicolegal protection of the examining physician rather than the victim. Further, the assumption which permeated many articles was that the victim was not an innocent party to the rape: was she seductive, sexually active, drinking, out at night, did she know the assailant? The burden of proof of innocence was the responsibility of the victim. Similarly, a spokesperson for a hospital noted "If she doesn't want to report it, she probably wasn't really raped."[2] While no such remark was made by either hospital or law enforcement in the previous case report, the existence of differing guidelines for treating the victim who does or does not wish to report or prosecute, and the combined behaviors of those personnel in connection with the child victim suggest the identical attitude.

In March of 1974, a bill that would have temporarily prevented the publication or broadcast of a rape victim's name was defeated by the House Judiciary Committee of the North Carolina General Assembly. A member of that committee warned that delaying publication of the victim's name for ten days would result in the filing of capricious charges. "So many women would scream rape if they knew they could hide behind the bill . . . I'm sitting in jail for ten days because some floozy has charged me with rape."[3] Similar attitudes are reflected in the criminal justice system with its stringent corroboration requirements in rape cases. In the case of Coltrane v. United States in 1969, the need for corroboration was supported because "we know from the lessons of the past that all too frequently such complainants have an urge to fantasize or even a motive to fabricate. . . ."[4]

The theme of woman as "fabricator" or liar pervades all of our attitudes about rape victims. The report of the District of Columbia Task Force on Rape concludes:[5]

> Victorian anachronisms appear to underlie many judicial decisions as well as the verdicts of even the most representative juries. These include the suspicion that a "proper" person should have absorbed substantial physical brutality to evidence lack of consent; that prior sexual experience of any kind is reasonable evidence of possible misconduct or "provocation" on the part of an unmarried victim; that "nice girls don't get raped and bad girls shouldn't complain."

This image of women as liars is likely the explanation for the assumption that women often make false charges of rape against men, even men they don't know. Law enforcement pesonnel are aware that false charges of crime do occur, but it is only in rape that it is assumed that the usual safeguards in the system are inadequate to protect the innocent from a lying witness. Contrast a charge of rape with that of robbery, where it is understood that property is taken from the victim without his/her consent, and there is no need to prove that fear of death or grave bodily harm was at issue. The contrast is striking if one imagines the cross-examination of a robbery victim which parallels that of the rape victim:[6]

> In other words, Mr. Smith, you were walking around the streets late at night in a suit that practically advertised the fact that you might be a good target for some easy money, isn't that so? I mean, if we didn't know better, Mr. Smith, we might even think that you were *asking* for this to happen, mightn't we?

The District of Columbia Task Force suggests that the law grants more protection to property than to the person, especially if the person is female.

Thus, medical institutions, law enforcement and the prosecutory system reflect the same mythology which society at large perpetuates about rape. Myths about rape are myriad and include the following:

1. The rapist is a sexually unfulfilled man carried away by a sudden uncontrollable urge.
2. Rapists are sick.
3. Rapists are strangers.
4. Rape occurs on the street, so as long as a woman stays home, she's safe.
5. Most rapes involve black men raping white women.
6. Women are raped because they ask for it by dressing seductively, walking provocatively, etc.
7. Only women with "bad" reputations are raped.
8. Most victims have been in trouble with the law in the past.
9. Only women in the lower social classes get raped.
10. Women can't be raped unless they want to be. A corollary of this might be that women enjoy rape.

Underlying much of this mythology is the notion that the victim is indeed "some floozy" whose testimony is likely to be malicious and deceitful. The victim, then, is considered a responsible and not an innocent party to the crime.

The most thorough study of the sociocultural aspects of rape was published by Amir[7] in 1971, with many of his findings documented by others as well. His data encompassed all cases of rape listed by the police in 1958 and 1960 in the city of Philadelphia, and does not include incest or statutory rape. His total sample included 646 victims and 1292 offenders.

Three quarters of the rapes involved one or two assailants (single rape—57%, pair rape—16%), with group rape (three or more assailants) the pattern in 27%. Of the total number of incidents, 71% were planned in advance, and only 16% could be considered explosive. Group rapes were planned in 90% of cases, and single rapes in 58% of cases. This serves to challenge the "uncontrollable urge" theory of rape. Although mythology has it that staying at home is safe, 56% of rapes occurred in the victim's residence, and the remainder were divided among automobiles, outside, and other indoor places. In only half of the cases was the rapist a stranger to the victim, while the remainder included casual acquaintances, neighbors, boyfriends, family friends, and relatives. Husbands are not included in these statistics because a sexual act between husband and wife is not usually considered a rape in this country.

In Hayman's study of 1,223 cases,[8] the age of the victim ranged from 15 months to 82 years, with 12% of victims under 12, 25% between 13 and 17, 32% between 18 and 24, and 30% over 25. The rapists were almost all under 30 years old. The overwhelming majority of reported rapes were *intra*racial, with rapist and victim of the same race. Most studies suggest a high proportion of black rape, but the significance of this is unclear. It is possible that black rapists are more likely to be reported and apprehended, while white rape may be grossly underreported or less aggressively pursued when reported. The Denver study[9] is an exception in that the percentage of victims by race was similar to the at large population, that is, 71% white, 15% black, 11% chicano.

In Amir's sample, physical force was present in 85% of cases, the remainder involving various degrees of nonphysical force, as coercion, and intimidation with or without weapons. Amir characterizes physical force in the following way:

Roughness (holding, pushing around)	29%
Nonbrutal beating (slapping)	25%

Brutal beating (slugging, kicking, beating by fists repeatedly)	20%
Choking and gagging	12%

Thus, in one third of the cases in which physical force occurred, extreme brutality was evident. A special comment should be made of group rape in which there is a higher frequency of both alcohol intake and prior criminal records, especially of sexual offenses. The assault is usually planned, and is more brutal in terms of beatings and subjecting the victim to sexually humiliating practices in addition to the rape. The above statistics do not include rape which ends in death. The media is replete with sensational examples, but this group is reported in the homicide statistics rather than as rape.

Victim behavior is described by Amir as submissive in 55%, with some degree of resistance in the remainder. Two factors which are always considered by the rape victim are the actual rape and the possibility of injury. At the time of the assault, the victim must decide whether she has a greater fear of the rape or of physical injury, and her actions will reflect her decision. This presents something of a bind in that resistance increases the victim's chance of escape, but also increases the likelihood of violence toward her should she not escape.

Perhaps the most significant finding of Amir's study is that rape occurs in a context of violence rather than passion: "Rape is a deviant act, not because of the sexual act per se, but rather in the mode of the act, which implies aggression, whereby the sexual factor supplies the motive." Similarly, Selkin[10] concludes that the purpose of rape is to humiliate and debase the victim, with the sex act itself secondary. While mythology has it that most rapists are sexually perverted monsters who hide in bushes and hate their mothers, Amir suggests that rapists are a danger to the community not because they are compulsive sex fiends but because they are violent and aggressive. They often appear psychiatrically normal, but tend to have criminal records of offenses against the person, of which rape is just one, and they usually commit the offenses with brutality and violence. In his series, rapists tend to be young members of lower class subcultures in which masculinity is expressed by displays of aggressiveness, which include sexual exploits of women. This is most evident in group rape, in which aggressive behavior is not the result of deviant sexuality, but of participation in a group which condones the use of force in attaining goals. Brownmiller[11] summarizes this theory of the "subculture of violence" as follows:

> Within the dominant value system of our culture there exists a subculture formed of those from the lower classes, the poor, the disenfranchised, the black, whose values often run counter to those of the dominant culture, the people in charge. The dominant culture can operate within the laws of civility because it has little need to resort to violence to get what it wants. The subculture, thwarted, inarticulate and angry, is quick to resort to violence; indeed, violence and physical aggression become a common way of life. Particularly for young males.

In recent years, there has been much written about rape as a symbol of the power relationship between the sexes:

> Rape is to women as lynching was to blacks: the ultimate physical threat by which all men keep all women in a state of psychological intimidation.
>
> SUSAN BROWNMILLER[11]

> Rape was an insurrectionary act. It delighted me . . .
>
> ELDRIDGE CLEAVER[12]

> It is always the woman who is raped. Rape is an aggressive act against women as woman. The rapist is educated to his behavior by his society. Rape is the extreme manifestation of approved activities in which one segment of society dominates another. Rape is a ritual of power.
>
> DEENA METZGER[13]

> Boy, it made me feel powerful.
>
> ALBERT DE SALVO[11]

> Rape is an act of aggression in which the victim is denied her self-determination. It is an act of violence which, if not actually followed by beatings or murder, nevertheless always carries with it the threat of death. And finally, rape is a form of mass terrorism, for the victims of rape are chosen indiscriminately but the propagandists for male supremacy broadcast that it is women who cause rape by being unchaste or in the wrong place at the wrong time—in essence, by behaving as though they were free.
>
> SUSAN GRIFFIN[14]

Weis and Borges[15] describe in some detail the process by which a person becomes a victim, and specifically, the way in which our cultural norms determine that women are "legitimate" victims for rape. Males learn that aggressive conquest is an acceptable substitute for "failures as a man in the economic and social status spheres"[7] so that aggressive and exploitive behavior towards women become part of their normative systems and they do not conceive of such behavior as wrong or a deviation from the normal. Some confirmation of this is seen in prison where rapists, unlike other sex offenders, rank high in the male prison hierarchy.[15] The rapist may justify his behavior with the logic used for the legitimation of victims in general: she was in some way already inferior, and she asked for and deserved it. Weis and Borges note that Amir himself uses this logic in his study. He describes a group of cases as "victim-precipitated rape" with the following characteristics: 92.7% involved violence; 45.9% of the victims were raped by more than one offender who humiliated the victim in 61.5% of cases; and the victim's behavior was resistive in 48.7% of cases. One might similarly imply that bank tellers precipitate bank robberies. While men are socialized to equate aggressiveness with masculinity,

women are brought up to think of themselves as sexual objects, subject to being acted upon by men. In this society, the relations between the sexes are seen as an instrumental exchange, whereby female servility is the price of male protection. The socialization process is such that women are educated to internalize the psychological characteristics of defenseless victims who have not learned or cannot apply the techniques of self defense and so must rely on the protection of others. In addition to what the woman learns about enacting the female role, she learns a mythology about rape that ensures a male advantage and provides the rationale for perceiving of her as a legitimate victim for rape. She is taught the cultural stereotype that the typical rape situation involves a stranger in a dark alley and that it is up to her to avoid both dangerous and compromising situations. In general, nice girls do not get into trouble or get raped.[15]

It seems appropriate to conclude this chapter with a statement by Susan Brownmiller, who has written with singular eloquence on the relationship between rape and power:[11]

A world without rapists would be a world in which women moved freely without fear of men. That *some* men rape provides a sufficient threat to keep all women in a constant state of intimidation forever conscious of the knowledge that the biological tool must be held in awe for it may turn to weapon with sudden swiftness born of harmful intent . . . men who commit rape have served in effect as frontline masculine shock troops, terrorist guerrillas in the longest sustained battle the world has ever known.

References

1. Hilberman, E.: Rape: A Crisis in Silence. *Psychiatric Opinion* publication pending.

2. Largen, M. A.: A Report on Rape in the Suburbs. Northern Virginia Chapter of National Organization of Women, p. 25, 1973.

3. Raleigh *News and Observer.* March 22, 1974.

4. *Coltrane* v. *United States*, 135 U.S. App. D.C. 298-99, 418 F. 2d 1131, 1134-35, 1969.

5. Report of the District of Columbia Task Force on Rape, Subcommittee of District of Columbia City Council, July, 1973.

6. Borkenhagen, C. K.: House of Delegates Redefines Death, Urges Redefinition of Rape, and Undoes the Houston Amendments. *American Bar Association Journal*, 61, 464-65, April 1975.

7. Amir, M.: *Patterns of Forcible Rape*. University of Chicago, 1971.

8. Hayman, C., Lanza, C.: Sexual Assault on Women and Girls. *American Journal of Obstetrics and Gynecology*, 103:3, 480-486, February, 1971.

9. Giacinti, T. A., Tjaden, C.: The Crime of Rape in Denver. Denver Anti Crime Council, 1973.

10. Summary and Recommendations of the National Rape Reduction Workshop. Denver Anti Crime Council, May, 1973.

11. Brownmiller, S.: *Against Our Will: Men, Women and Rape*. Simon and Schuster, New York, New York, 1975.

12. Cleaver, E.: *Soul on Ice*. 1962.

13. Metzger, D.: It is Always The Woman Who is Raped. Presented at Special Session on Rape, APA, Anaheim, California, 1975. Publication pending, American Journal of Psychiatry, April, 1976.

14. Griffin, S.: Rape: The All-American Crime. *Ramparts*. September, 1971.

15. Weis, K., Borges, S. S.: Victimology and Rape: The Case of the Legitimate Victim. *Issues in Criminology*, 8:2, 71-115, Fall, 1973.

CHAPTER

TWO

LEGAL ASPECTS OF RAPE

According to the Uniform Crime Reports released by the FBI,[1] 46,430 women were victims of rape in 1972. This volume represents an 11% increase over 1971, and a 70% increase over 1967. In 1974, there was an estimated total of 55,210 forcible rapes, an 8% increase over the preceding year, and a 49% increase over 1969. FBI comparative statistics confirm that rape is the fastest growing of the violent crimes, with almost a third of the total volume reported in the southern states. While better reporting may account for part of the increase, it is felt that these statistics greatly underreflect the actual incidence of rape. Given the climate of opinion about rape, it is not surprising that forcible rape is probably one of the most underreported crimes in the United States today, with educated estimates that between 50 to 90% of rape cases go unreported. The FBI attributes underreporting to "fear and/or embarrassment on the part of the victims."[1] She is afraid of public accusation of provocation or active participation in the rape. She is fearful of the reactions of those close to her, whether husband, boyfriend, parents, or friends. In the case of a young victim, the parents wish to protect the child from the publicity and the legal ordeal. If the assailant is a close friend, relative, or employer, there are additional pressures not to report.

The natural channels for reportage are hospitals and law enforcement agencies, a common feature being that most citizens trust neither one of these institutions to deal with rape. Hospitals suffer from lack of personnel trained to work with victims both in the crisis period and in follow-up, and from lack of consistent and clear procedures for evidence collection. In the absence of a formal policy for the treatment of victims in crisis, personal attitudes and fears prevail on the part of the staff, both with regard to the victim and the criminal justice system. A clinician, then, might choose to disbelieve the victim rather than to face the prospect of a court appearance. Similarly, law enforcement also suffers from lack of personnel identified and trained to work with rape victims. The victim who reports does not receive consistent treatment because of high rates of police turnover, rotating shifts, and personal attitudes. The public equates reportage with prosecution, and is fearful of harassment by law enforcement to prosecute. Finally, reportage and prosecution are both equated with the victim's life and past being made a matter of public record.

Unfortunately, there is considerable reality to the victim's fears, as reflected in the attitudes about the victim which prevail in the criminal justice system and in the statutes about rape. The report of the Center for Women Policy Studies[2] comments that "the credibility of the rape victim is questioned more than that of any other victim of crime. This extraordinary concern for the authenticity of rape complaints is manifested in the FBI's Uniform Crime Reports—forcible rape is the only crime for which an 'unfounded' rate is calculated." It is also the only violent crime for which corroboration is required. Corroboration is defined[3,4] as support of a fact by evidence independent of the mere assertion of the fact. It is not considered "supporting evidence that can be used to strengthen the prosecutor's case, it is evidence that is *required* to prove that rape occurred."[4] The victim's testimony, then, is insufficient grounds for conviction even if her testimony is probable and consistent.

One of the consequences of the corroboration requirement is that it is the victim who stands trial. The assumption that women will make false accusations against men "makes the victim's testimony the central object of inquiry and not the rape incident itself."[4] This assumption is pervasive in the literature on the law, and is legitimized by the use of pseudopsychological jargon, as in the following statements:

> No judge should ever let a sex-offense charge go to the jury unless the female complainant's social history and mental makeup have been examined and testified to by a qualified physician.[5]

> Today it is *unanimously* held (and we say "unanimously" advisedly) by experienced psychiatrists that the complainant woman in a sex offense should always be examined by competent experts to ascertain whether she suffers from some mental or moral delusion or tendency,

frequently found especially in young girls, causing distortion of the imagination in sex cases.[6]

Women frequently have fantasies of being raped. Dr. Karl A. Menninger has said that such fantasies might almost be said to be universal. And in a hysterical female, these fantasies are all too easily translated into actual belief and memory falsification. It is fairly certain that many innocent men have gone to jail on the plausible tale of some innocent looking girl because the orthodox rules of evidence (and the chivalry of judges unversed in psychiatry) did not permit adequate probing of her veracity.[7]

Judge Hale's often quoted remark that rape "is an accusation easily to be made and hard to be proved, and harder to be defended by the party accused, tho never so innocent" is a remarkable distortion of the victim's experiences in the criminal justice system. The laws pertaining to rape make it highly unlikely that the victim will even report to law enforcement much less agree to the ordeal of the trial. Wood[8] summarizes the complainant's ordeal throughout the various phases of the criminal justice procedure, which includes the initial report to the police, the investigation, the grand jury and the trial itself. At best, the initial report becomes a traumatic event because of the need to give an exquisitely detailed history of the rape, which must be repeated to a variety of personnel connected with the subsequent investigation. Although there has been recent attention focused on the importance of law enforcement sensitivity to the victim's mental state and the management of victims in crisis,[2,9] this is not the prevailing concern of most law enforcement agencies. Victim treatment may be impersonal and unsupportive if not frankly disbelieving and hostile. Thus, the victim can expect questions about how many orgasms she had during the rape, what fantasies she had while he was doing it, and allusions to past sexual "misconduct." Wood points out that the police have considerable discretion as to whether or not any action is taken to obtain a conviction; they may refuse to accept the complainant's charges, or neglect to work on cases they feel are unsubstantiated. The prevalent rape mythology makes it likely that law enforcement personnel will infer consent when none is given or assume victim precipitation.

Assuming that the complainant has survived the investigation and the grand jury has found probable cause that the defendant committed the crime, the stage is set for the trial, wherein the courtroom becomes the microcosm of society in which every stereotyped assumption about women is enacted, and one sees

> two different facets of the female stereotype in the mythology of male supremacy. On the one hand, the female is viewed as a pure, delicate and vulnerable creature who must be protected from exposure to immoral influences; and on the other, as a brazen temptress, from whose seductive blandishments the innocent male must be protected. Every woman is either Eve or Little Eva—and either way, she loses.[10]

There are three major elements of the legal defense in a rape case, as delineated by Hibey[3] and Connell:[4]

1. identification, that the man accused is the perpetrator of the crime.
2. penetration, that a sexual act took place.
3. lack of consent, that intercourse was not a voluntary act on the part of the woman.

Corroboration is acquired through lengthy interviews with the complainant, whose motivation to falsify is under close scrutiny. Independent corroboration may be in the form of an eye-witness, evidence acquired at the crime scene, and from the body of the victim or defendant. Circumstantial evidence of penetration rather than direct eyewitness evidence is usually the case. Hibey[3] summarizes the circumstantial evidence of penetration to include medical evidence and testimony, evidence of breaking and entering the complainant's residence, condition of clothing, bruises and scratches, emotional condition of complainant, opportunity of the accused, conduct of the accused at the time of arrest, presence of semen or blood in the clothing of the accused or the victim, promptness of victim's report to the police, and lack of motive to falsify. Hair and sperm are both subject to typing so that some sources of corroboration for identification also serve to establish proof of corroboration for penetration. The complainant who instinctively bathes and changes clothes after the rape, or who is too frightened to report immediately, or who presents a calm exterior may not be believed. The defense counsel can attack the corroboration of identification/penetration by introducing evidence of mental illness, previous consensual intercourse, previous false accusations of rape, or an unchaste reputation.

Consent is usually the issue on which court cases hinge. Most laws demand a high degree of resistance from the victim, based on the belief that a healthy woman cannot be forcibly raped. Wood[8] reports one of many cases which are lost because of insufficient resistance by the victim. She describes an assault in which the assailant sodomized two students and admitted the crime. One woman was beaten repeatedly on the head, and the other had ten sizable bruises on her body. The jury acquitted because the women didn't resist enough or try to escape. A growing number of statutes appear to recognize that the amount of resistance depends on the circumstances of the attack and the implications of continued resistance so that victim compliance is not a sign of consent but of futility. The report of the Center for Women Policy Studies[11] comments on this issue:

> When and how a woman should resist a rapist are under hot debate. . . . The few published studies of convicted rapists indicate that there are three or four different categories of offenders whose motives, methods, and reactions to resistance differ. Since rapists do not wear identifying labels, a woman cannot know which type she is confronting. . . .
>
> An important element of any program on defense against rape should be emphasis of the right of the woman to submit. Although some people successfully resist robbery, others are killed in the attempt, so no one is counseled to fight a robber. Although there are different values at stake, the choice should still be the victim's. Even a person who wants to resist and is trained to fight may be unable to do so when confronted with a situation which she or he perceives as dangerous.

Consent is considered an affirmative defense in which the occurrence of the event is not disputed, but legitimized or excused.[3] Since the success of the consent defense depends on which version of the facts the jury will believe, the defense must not only make their case believable, but must also make the other side unbelievable. This is done largely through evidence of the complainant's general character or reputation for unchastity, the implication being that prior consensual intercourse whether with the defendant or someone else denies the woman her right to choose her sexual partner.

Many women feel that the post-rape trauma is simply not worth it when there is little reason to believe that the assailant will be punished for his crime. Not only is rape the fastest growing of the Index crimes against the person, but among these it has the lowest proportion of cases closed by reason of arrest.[2] In 1974, only half of the reported rapes led to an arrest, and 40% of men arrested were never prosecuted. Of the remaining 60% who were prosecuted, half were acquitted or had cases dismissed.[1] In Denver, for example, 950 cases were reported to the police in 1971-72, and only 41 cases were brought to trial.[12] Gates[13] cites a 1966 study of jury trials to see if judges would have made the same verdicts as juries, and if not, what factors the judges thought influenced juries. In rape cases in which there was no extrinsic violence, one assailant, and no prior acquaintance of the victim and the assailant, the judge and jury would have reached the same conclusion in only 40% of cases. In the remaining 60%, the judge would have convicted where the jury acquitted. The judges concluded that in the absence of external evidence of violence, jurors ascribe to the complainant some contributory or precipitant behavior. On the basis of jury decisions, Hibey suggests that the

> jury's assessment of the credibility of the witnesses and their evidence is not always rational. This phenomenon stems in large part from certain ideas jurors have about the crime of rape, some of which are believed with such ferocity that jury verdicts are often examples of outright nullification—the ultimate and extreme exercise of the fact finder's prerogative.

It is not surprising that these beliefs include the ideas that black men are more likely to attack white women than black women, that a hitchhiking woman implies consent, that a woman who sleeps in the nude is looking for intercourse, and that a victim who doesn't wear undergarments was probably not raped.[3] Jurors represent the mainstream of a society in which women are condemned for their freedom.

In addition to the complex social forces deterring reports of rape, the victim has a set of internalized beliefs stemming from her own socialization which decrease the likelihood of reporting. She is taught the stereotyped view that the rapist is a sex-starved monster or pervert who is waiting in some dark alley. When the reality of the rape does not coincide with this view, as it rarely does (half of the assailants are acquaintances if not trusted friends), she too is left to wonder about her own complicity in the event. She also "knows" that nice girls don't get raped and that it is not possible to rape a woman against her will, and this further contributes to her silence.

The Prince Georges County Task Force on Rape comments on the silence of the victim as follows:[14]

> Rape is a serious crime of assault on the body, but more grievously on the psyche of a woman. All too often, she is treated at best as an object, a piece of evidence, and is made to relive the experience, must face the incredulity of the police, the impersonality of the hospital, and then must defend herself in court. Having been socialized to be passive, she is nevertheless expected to have put up a battle against her attacker. Her previous sexual experience can be used to impute her instability though the defendant's background often cannot be brought up against him. She does not have the benefit of a retained lawyer and sometimes the prosecutor does not have the time or perhaps the insight to prepare her beforehand for the ordeal of the trial. She suffers serious psychological stress afterwards, largely due to the guilt and shame imposed by society. She may not recognize a need for professional help or she simply cannot afford it.

Thus, there are multiple forces which act as deterrents to reporting. One can assume that it will be necessary to create a sympathetic climate which allows the victim to identify herself as such to the hospital, law enforcement, and judicial systems. The House of Delegates of the American Bar Association has recently adopted a resolution[15] which would protect the victim from unnecessary invasion of privacy and the consequent psychological trauma. The proposed changes include the elimination of corroboration requirements which exceed those applicable to other assaults, revision of the rules of evidence relating to cross-examination of the complaining witness, development of new procedures for police and prosecution in processing rape cases, and establishment of rape treatment and study centers to aid both the victim and the offender. Gates[13] and Wood[8] both recommend a redefinition of rape to include oral and anal intercourse and a delineation of varieties of rape to be contingent on the degree of force or violence used. They suggest that the resistance standard be dependent on "reasonable fear" and whether victim resistance was reasonable to expect under the circumstances of the assault. Reduced penalties for rape, in line with comparable violent crimes, would increase the likelihood of jury convictions. The impact of such changes would eliminate the present distinction between victims of rape and victims of other crimes.

References

1. *Uniform Crime Reports for the United States.* Prepared by the Federal Bureau of Investigation, Washington, D.C., 1972 and 1974.

2. Rape and its Victims: A Report for Citizens, Health Facilities and Criminal Justice Agencies, *The Police Response: A Handbook.* Prepared by the Center for Women Policy Studies, Washington, D.C., 1975. Copies available from Law Enforcement Assistance Administration, Washington, D.C.

3. Hibey, R. A.: The Trial of a Rape Case: An Advocate's Analysis of Corroboration, Consent, and Character. Reprinted in *Rape Victimology*, ed. Leroy G. Schultz, Charles C. Thomas, Springfield, Ill., 1975 (originally published in *American Criminal Law Review*, 11:2 Winter, 1973).

4. New York Radical Feminists. *Rape: The First Sourcebook for Women*, N. Connell and C. Wilson eds., New American Library, New York, New York, 1974.

5. Wigmore, J.: *Evidence*, Vol. 3, Sec. 924 (a), 3rd Ed., 1940.

6. 1937-38 Report of the American Bar Association Committee on the Improvement of the Law of Evidence.

7. Guttmacher, M., Weinhofen, H.: *Psychiatry and the Law*, 1952.

8. Wood, P. L.: The Victim in a Forcible Rape Case: A Feminist View. Reprinted in *Rape Victimology*, ed. Leroy G. Schultz, Charles C. Thomas. Springfield, Ill., 1975 (originally published in *American Criminal Law Review*, 11:2 Winter 1973).

9. Bard, M., Ellison, K.: Crisis Intervention and Investigation of Forcible Rape. *The Police Chief*, 41:5, 68-73, May, 1974.

10. Johnston, J. D., Jr., Knapp, C. L.: Sex Discrimination by Law: A Study in Judicial Perspective. *New York University Law Review*, 46:4, 704-705, October, 1971.

11. Rape and its Victims: A Report for Citizens, Health Facilities and Criminal Justice Agencies, *The Response of Citizens' Action Groups: A Handbook.* Prepared by Center for Women Policy Studies, Washington, D.C., 1975. Copies available from Law Enforcement Assistance Administration, Washington, D.C.

12. Giacinti, T. A., Tjaden, C.: The Crime of Rape in Denver. Denver Anti Crime Council, 1973.

13. Gates, M. J.: The Rape Victim on Trial. Presented at Special Session on Rape, APA, Anaheim, California, 1975.

14. Report of the Task Force to Study the Treatment of the Victims of Sexual Assault. Task Force of the County Council of Prince George's County, Maryland, March, 1973.

15. *American Bar Association Journal.* 61, 464-65, April, 1975.

CHAPTER

THREE

MEDICAL ASPECTS OF RAPE

The physician treating the rape victim has the following tasks:[1]

1. immediate care of physical injuries
2. prevention of venereal disease
3. prevention of pregnancy
4. proper medicolegal examination with documentation by evidence collection for law enforcement
5. prevention or alleviation of permanent psychological damage

Since the issue of psychological trauma is dealt with elsewhere in this monograph, this section will focus on the medicolegal aspects of the victim's care. The psychiatrist or counselor needs familiarity with medical aspects of rape, since the victim's physical status presents yet another dilemma during the crisis period.

Traumatic Injuries
Although it is said that most victims do not suffer serious physical, injury, most case reports describe traumatic injury in a significant number of instances. Hayman[2] describes a sample of 2,190 victims, in which 82 sustained severe physical injury noted at the time of the initial medical examination. Of the 24 who were hospitalized, 6 children and 1 adult had vaginal or vaginoperineal tears, while 1 child and 16 adults were admitted for other injuries. The child had multiple head injuries, and the adults sustained a variety of injuries which included abrasions, lacerations, stab wounds, fractures, and a torn digital nerve. The remaining 58 victims required major treatment in the emergency room, of whom 11 were children with vaginal or vaginoperineal lacerations. Many hundreds more required treatment of minor injuries. During Hayman's study period, there were 4 proven murders from physical assaults which included

rape, and 3 murders in which rape was suspected. In Massey's series of 501 cases,[1] 51 showed external evidence of trauma, ranging from small cuts to severe contusions and a facial fracture. Two were victims of prolonged kidnapping as well as rape and had evidence of multiple episodes of physical trauma. Gynecologic injuries were evident in 25 victims, spanning a spectrum of laceration of the hymen to rupture of the cul de sac. Complete vaginal penetration in children frequently resulted in perineal lacerations. In Burgess and Holmstrom's sample of 146 victims,[3] 86 of the women had visible bruises, either from a weapon or as the result of being hit with the assailant's hand or fist. The head and neck area received the highest proportion of trauma, and 21 of the victims required medical, surgical, or orthopedic consultation. 57 victims had signs of injury on gynecological examination.

Prevention of Venereal Disease

Hayman[2] reports documentation of preexisting venereal disease during the initial examination by 94 positive serologic tests and 54 positive smears for gonorrhea. In the same sample, 5 victims, one of whom was a four year old child, developed syphilis; one victim developed lymphogranuloma and 76 developed gonorrhea. Medical treatment of victims should routinely include venereal disease prophylaxis, either with intramuscular penicillin or another antibiotic if there is a history of penicillin allergy. The recommended dose of penicillin is curative for gonorrhea as well as for syphilis contracted at the time of the assault. It is not curative for a previously established syphilis infection, so that victims with positive serologic tests need follow-up treatment.

Prevention of Pregnancy

In Hayman's sample, 39 victims were pregnant at the time of the rape, and 13 became pregnant as a result of the rape. Fears of pregnancy are frequent concerns of victims, and the possibility of pregnancy must always be considered as part of the medical evaluation of rape victims. Information about menstrual status, current place in the menstrual cycle, current birth control practices, a possible preexisting pregnancy, and feelings about abortion are all considered in making the appropriate disposition. Pre- and postmenopausal females as well as those currently using oral contraceptives or intrauterine devices need no further evaluation. The victim who is unprotected and in midcycle has two available options. The administration of large doses of estrogens ("morning after treatment") within 5 days of sexual contact appears to be effective in preventing implantation, with diethylstilbestrol most widely used. Side-effects are prominent, with half of the patients experiencing nausea, for which antiemetics are necessary. In one follow-up survey of the side-effects of estrogen treatment,[1] 10% of the respondents did not complete the drug treatment in part because of the side-effects. In addition to treating a victim in crisis with a drug which induces nausea and vomiting for the next few days, there have been grave questions raised about the long-term effects of diethylstilbestrol.[4] Because of these issues, it is likely that in the future Premarin will assume increasing importance in the hormonal prevention of pregnancy, but this drug also may have long-term consequences.

The option preferred by an increasing number of women would be to wait until the next expected menstrual period, and to perform a menstrual extraction if the period does not occur. These alternatives and their implications, however are not usually presented openly to the victim, most physicians favoring estrogen treatment.

Obviously, the victim who is vehemently against abortion on religious or other grounds would appropriately be treated with estrogens. In all cases, the victim does need reassurance that pregnancy prevention or pregnancy interruption will be accomplished, and that no victim need carry a pregnancy to term.

Medicolegal examination
The purpose of the medicolegal examination is to collect evidence which serves to document corroboration of identification as well as of the assault. It is not the responsibility of the physician to decide whether or not a rape occurred, but simply to gather the evidence which will subsequently allow the courts to make this judgment. The evidence which the physician collects is likely to be the only evidence the victim has, in the absence of an eye-witness, and therefore the examination must be done with painstaking thoroughness. The ACOG Technical Bulletin on this subject is an excellent resource.[5] The following outlines accepted procedure:

1. An interview is required to assess the psychological and

physical status of the victim. This must include information about her menstrual status, and specific details about the rape incident so that the physician has a guide to which parts of the body need especially close scrutiny for bruises and trauma. Most guidelines stress a description of mental state, which may be less for the benefit of the victim than as "proof" of possible fabrication. Thus, one recent article suggests that "a lack of emotional response . . . might indicate that the victim is fabricating the complaint."[6] Information about the victim's behavior between the time of the rape and the examination is necessary to find out whether the victim bathed or changed clothes since this obviously affects the evidence.

2. Description of wearing apparel (soiled, torn, bloody clothing should be saved as evidence).

3. Physical examination

a) General appearance: bruises, lacerations, evidence of trauma, with type and distribution.

b) Foreign material (dirt, vegetation): material collected from buttocks, vulva, and other parts of body should be saved.

c) External genitalia: evidence of trauma.

d) Vaginal examination with a nonlubricated, water moistened speculum for further evidence of trauma, seminal fluid, collection of laboratory specimens.

4. Laboratory Examination

The following listing is described in the context of the genital area. Where there is a history of oral or rectal penetration, specimens for laboratory examination must obviously include these additional areas.

a) Pubic hair combings *and* pubic hairs pulled from the victim: hair is also subject to typing, which can serve to corroborate identification and sexual contact.

b) Air-dried smear to serve as permanent record of presence or absence of sperm.

c) Swab from vaginal pool to be examined for acid phosphatase and semen typing. Semen is typed as is blood and may or may not corroborate the victim's identification of the assailant. Acid phosphatase remains elevated for up to 12 hours after intercourse, and with the increase in vasectomies, becomes an important test.

    d) Saline suspension from vaginal pool to determine presence or absence of motile or nonmotile sperm.

    e) Smears and cultures for neisseria.

    f) Scrapings from under finger nails may also assist in identification of the assailant.

5. Photographs can be taken to document the extent of trauma.

6. Chain of evidence.

For legal reasons, all specimens must be carefully tagged and clearly identified, with the names of both the victim and each person having custody of the evidence at each stage of handling specimens. Unless this is accomplished, the prosecutor may have difficulty in submitting the reports in evidence.

7. Finally, it is recommended that the full medicolegal examination be performed regardless of whether or not the victim plans to prosecute. The physician is likely to see the victim in the immediate aftermath of the rape when the victim has not yet made a decision about criminal justice involvement. Some programs save evidence for up to six months, after which it is discarded.[7] If the evidence is not collected, the possibility of prosecution is nonexistent.

References

1. Massey, J. B., Garcia, C. R., Emich, J. P.: Management of Sexually Assaulted Females. *Obstetrics and Gynecology*, 38:1, 29-36, July 1971.

2. Hayman, C. R., Lanza, C.: Sexual Assault on Women and Girls. *American Journal of Obstetrics and Gynecology*, 109:3, 480-486, February 1971.

3. Burgess, A. W., Holmstrom, L. L.: *Rape: Victims of Crisis*. Bowie, Maryland, Robert J. Brady Co., 1974.

4. Weiss, K.: *What the Rape Victim Should Know About the "Morning-After Pill,"* Advocates for Medical Information, 2120 Bissonnett, Houston, Texas 77005, 1975.

5. *Suspected Rape*. ACOG Technical Bulletin Number 14, July 1970 (Revised, April, 1972).

6. Enos, W. F., Beyer, J. C., Mann, G. T.: The Medical Examination of Cases of Rape. Chapter 12, *Rape Victimology*, ed. LeRoy G. Schultz, Charles C. Thomas, Springfield, Illinois, 1975 (originally published in *Journal of Forensic Sciences*, 17:1, 1972).

7. Emergency Room Rape Crisis Program. North Carolina Memorial Hospital, Chapel Hill, North Carolina.

CHAPTER
FOUR

THE RAPE VICTIM IN THE HOSPITAL

This monograph has already addressed itself to the issue of professional attitudes about rape, and the ways in which attitudes affect behavior, whether in the emergency room or the courtroom. The purpose of this section is to describe some of the specific problems inherent in the hospital treatment of the victim and to provide a set of guidelines which can be utilized in instituting crisis programs for victims of sexual assault in a hospital setting.

The Report of the District of Columbia Task Force on Rape[1] describes many of the problems which currently prevail in hospital treatment of victims. They suggest that "many doctors do not want to examine a rape victim because they do not wish to be called to testify. Some doctors who examine victims falsify medical reports for court, minimizing or neglecting entirely signs of trauma in an attempt to avoid being called in to testify." Further, physicians usually have had no special training in the treatment of the physical and emotional trauma resulting from the rape or in the methods of evidence collection. Hospital policies vary widely, with examinations performed by gynecologists in some institutions, while at others the lowest ranking physician without training in gynecology does the examination. Medical treatment is often inadequate and psychological treatment usually nonexistent. Victims have been known to wait for up to five hours before receiving medical attention, which is often brusque and impersonal. Issues concerning pregnancy and venereal disease may not be dealt with at all. There may be no formal procedure for collection of evidence, and even when such procedure exists, it may be by-passed. Issues of confidentiality become critical either because hospital policy dictates that parents give consent before a minor is treated or because some hospitals automatically call the police whether a victim wishes to report or not. Some victims, then, choose to forgo treatment rather than have parents or police know of the rape.

There are currently a variety of formal reports[1-4] which make recommendations for improving hospital services to rape victims. These include the following:

1. At least one medical facility in any given community should have a formal program for the comprehensive treatment of rape victims, with such services available on a 24-hour basis. Both law enforcement and the public should be made aware that these services exist. Those facilities which do not have comprehensive programs should have, as a minimum, a set of guidelines for treating rape victims.

2. Since treatment of the rape victim involves an interface between medical and legal issues, programs designed by medical facilities should be part of a community-oriented approach, in which there is a cooperative effort by hospitals, citizen groups, law enforcement, and prosecutory agencies.

3. Treatment of the rape victim should be given high priority, second only to life-threatening illnesses or accidents, and should not be contingent on cooperation with the criminal justice system.

4. All services offered to the victim must be based on informed consent. The victim has the right to information about all available options and thereby retains control throughout the hospital process.

5. Adequate care of the victim is facilitated by a team treatment model. The team should include a support person, nurse, physician, and appropriate consultants, among them mental health professionals. The support person serves as an advocate/counselor/guide for the victim throughout the hospital process, and must have familiarity with both crisis intervention techniques and criminal justice procedure. The nurse, usually a member of the regular emergency room staff, has a role in coordinating the total treatment plan, assessing needs, and explaining procedures. The physician has primarily a medicolegal role. While there are varying opinions about whether the physician ought to be a gynecologist, family practitioner, or specialist in trauma, there is consensus that an experienced clinician rather than a novice assume responsibility for victim care. An experienced clinician is also likely to be a more permanent member of the hospital staff, and therefore more easily accessible in the event of a court appearance one or two

years later. Because the physician's medicolegal role makes it highly probable that the victim will perceive him/her as an investigator rather than an ally, counseling responsibilities are best delegated to the other teams members. The role of the psychiatrist as a team consultant is discussed in Chapter 9. Finally, while it is desirable to be able to offer the victim a choice regarding the sex of the rape team personnel, sensitivity and experience may be more important determinants of a successful intervention.

6. All hospital personnel who will have contact with the victim must be screened and educated so that rape victims are not exposed to additional insult. The training program for this group of personnel should include crisis intervention theory and practice, sensitization to the physical and emotional trauma of rape, and medicolegal issues. It is desirable that training efforts be accomplished in collaboration with citizens' groups and local criminal justice agencies.

7. Hospitals must have clear procedural guidelines for victim care and evidence collection, with a specific and unambiguous description of the role and function of each member of the rape crisis team. The medicolegal examination is facilitated by prepackaged "rape kits" which contain all of the information and equipment necessary for examination and evidence collection.

8. Crisis intervention should be immediately available to all victims and their families/significant others at the time of initial contact with the hospital. This is discussed in detail in Chapters 6 and 7.

9. The victim should be informed both verbally and in writing about all treatment procedures performed in the emergency room, with special emphasis on those steps taken to prevent pregnancy and venereal disease. It is assumed that the victim is experiencing considerable confusion at the time of the hospital visit, and written information serves as reassurance that the appropriate preventive measures have been taken.

10. Follow-up treatment is complicated by the profound denial which often follows in the immediate aftermath of the rape. The victim attempts to "forget" the assault, and may not keep subsequent appointments. For this reason, it becomes especially important to provide all available services during the initial crisis contact since this may be the only intervention for a given victim. Follow-up

efforts may be more successful if they involve contact with the original treatment team members rather than new institutions and personnel. Thus, the team physician might arrange to see the victim herself, and the support person can maintain continuing contact. When formal mental health referrals are indicated, the need for such services must be explained so that the victim does not assume that because she was raped, she must be mentally ill. Obviously, all follow-up treatment plans are contingent on the victim's informed choice, and all referral agencies must be sensitized to the issues surrounding rape.

11. The issue of financial responsibility for the rape victim's medicolegal examination has been the object of recent attention. While some states have statutes providing for financial compensation to victims of violent crimes, rape victims have not been considered as appropriate beneficiaries of such laws because rape is traditionally viewed as a sexual rather than a violent crime. The victim is expected to go through the ordeal of reporting to hospital and law enforcement and then to face the trauma of the trial. It seems inappropriate that she also be expected to absorb the cost of the state's evidence. Payment for hospital services rendered should not be the responsibility of the victim, but should come from public funds.

12. While interagency collaboration is necessary to provide comprehensive services to the victim, all efforts must be made to protect the confidentiality of the victim. During the course of the examination, information about the victim may surface which is relevant to medical treatment, but irrelevant and potentially damaging in court, for example, pre-existing venereal disease or contraceptive usage in a single woman. Team members must exert some caution in what is entered into the victim's medical record, some hospitals using a satellite record for information which may be abused in court. Although treatment is not contingent on reporting to police, it is desirable for law enforcement to know about the incidence of rape. An anonymous or "blind" report might be instituted, in which the location and *modus operandi* of the assailant are described, but the name of the victim withheld.

There are at present a growing number of hospital-based rape crisis programs which have incorporated these recommendations. While each program has unique features based on the nature of the catchment area, availability of community resources, state law, and local criminal justice procedure, there are striking commonalities to all of these programs. McCombie *et. al.*[4] describe the process of program development which is invariably one of extraordinary complexity, in that training and collaborative efforts involve a large and interdisciplinary group of personnel in the hospital, the community, and the criminal justice system. Clarification of the roles of all personnel who are in contact with the victim appears to have a major role in decreasing staff anxiety, so that one's attitudes and biases are less likely to be acted out at the victim's expense. Thus, clear and unambiguous guidelines which define the roles and functions of the nurse, physician and counselor are essential to successful treatment.[5-7]

One such program[8] emphasizes a team approach to the care of the victim in the emergency room. The team consists of the emergency room nurse, gynecologist, and counselor. Hospital security is responsible for maintaining the chain of evidence and collaboration with law enforcement personnel who may have brought the victim to the emergency room. As soon as the victim enters the emergency room area, she is removed to a private room away from the public waiting area. If she is accompanied by family or friends, they are allowed to remain with her in the emergency room, if she desires. The emergency room nurse is the first member of the team to see the victim, and it is her responsibility to briefly assess the victim's status, and to determine the treatment priorities. Whether the counselor or the gynecologist is called first depends on the relative importance of emotional and physical trauma. If she is too distraught to be examined, the counselor is called first. If she has severe physical injuries requiring immediate attention, the gynecologist is called. The nurse also has responsibility for preparing the victim for the emergency room experience, with careful explanations of what will happen, and reassurance that nothing will happen without the victim's understanding, consent, and readiness.

The gynecologist has primary responsibility for a careful history, physical examination, treatment of injuries, prevention of pregnancy and venereal disease, and collection of evidence. Protocol for evidence collection has been standardized by clearly written guidelines[6,7] and evidence kits.

The counselor, who is a mental health professional with special training in the medical, legal, social and psychological aspects of rape, has a variety of functions which include:

1. Crisis intervention: helping the victim and her family/friends deal with the emotional trauma of the rape event.

2. Information: the medical and legal ramifications of rape, the implications and specifics of reporting, prosecuting, etc. The counselor also files a blind report directly. Much of this information is contained in a booklet about rape which is given to every victim.[9]

3. Referrals: resources for counseling, medical needs, criminal justice information after she leaves the emergency room. Victim follow-up includes phone contact by the counselor within the next 24-48 hours, and a gynecology appointment two weeks later, at which time an assessment of the victim's emotional state is made, and brief counseling is available.

There is a supervisory meeting conducted by a psychiatrist and a social worker for individual supervision of each case, with attendance by the entire team. Counselors meet at monthly intervals for group discussion of all crisis interventions. Program administration is handled jointly by the emergency room and the department of psychiatry, with a psychiatrist serving as medical director. The interdisciplinary coordinating committee for the program consists of some 8 members, each of whom assumes responsibility for a discrete area, as liaison to law enforcement or community-based rape crisis program or public information and media. This program is not funded by any external sources. The mental health professionals, which include psychiatrists, psychologists, social workers and nurses, are all members of the hospital staff who have volunteered their time to take night and weekend call. The hospital has absorbed the costs for maintaining the program, which serves as a model for an effective and relatively inexpensive way of providing adequate treatment for victims of rape.

References

1. Report of the District of Columbia Task Force on Rape. District of Columbia City Council, July 1973.

2. Report of the Task Force to Study the Treatment of Victims of Sexual Assault. County Council of Prince George's County, Maryland, March 1973.

3. Rape and Its Victims: A Report for Citizens, Health Facilities and Criminal Justice Agencies, *The Response of Medical Facilities: A Handbook.*

Prepared by Center for Women Policy Studies, Washington, D.C., 1975. Copies available from Law Enforcement Assistance Administration, Washington, D.C.

4. McCombie, S. L., Bassuk, E., Savitz, R., Pell, S.: Development of a Medical Center Rape Crisis Intervention Program. Presented at Special Session on Rape, APA, Anaheim, California, 1975. Publication pending, American Journal of Psychiatry, April, 1976.

5. Guidelines for the Treatment of Suspected Rape Victims. Chicago Hospital Council, February 1974.

6. Guidelines for Care of Victims of Rape and Sexual Assault. Emergency Room Rape Crisis Program, North Carolina Memorial Hospital, Chapel Hill, North Carolina, 1975, (see Appendix II).

7. Talbert, L. M., Warren, D. G.: Guidelines for Management of Suspected Rape. Chapel Hill, North Carolina, 1974, (see Appendix I).

8. Emergency Room Rape Crisis Program. North Carolina Memorial Hospital. Chapel Hill, North Carolina.

9. Boyles, J., Cole, K., Donadio, B., Hilberman, E., Peace, J., Reice, T.: *Information and Guidance for Adult Victims of Rape.* Prepared for the North Carolina Memorial Hospital Emergency Room Rape Crisis Program, 1974, (see Appendices III and IV).

CHAPTER

FIVE

COMMUNITY RAPE CRISIS CENTERS

In an historical context, it has been the women themselves who first identified rape as a traumatic event. They have taken the initiative in sensitizing our medical, social, and legal institutions to the extent to which cultural biases and attitudes have affected the treatment of the victim. Women in groups, first as informal groups of friends and later as formal organizations, have provided the stimulus for changes in hospital and criminal justice procedures as well as changes in the law itself. They have taken concrete action to provide support to victims and their families at a time when few if any professional resources were available. That this monograph has been written by a physician is an indication of their success.

The last decade has witnessed the spontaneous appearance of growing numbers of community-based rape crisis centers as part of a nation-wide anti-rape movement.[1-5] These centers are largely staffed by volunteer nonprofessional women, some of whom have been raped in the past, or who have been close to someone who was raped. In keeping with the philosophy that "women need to organize themselves to help each other in a male-dominated culture which is insensitive to women's needs,"[5] men are not usually accepted as volunteers in rape crisis programs. In some areas, however, men have been recruited to work with the victim's significant other males, or have organized themselves separately to support the sister group and to provide services to male friends and relatives of rape victims.

Most community rape crisis centers have similar goals which include the following:[5]

1. To provide supportive services to victims.
2. To reform the institutions which deal with victims.
3. To educate themselves and the public on rape-related issues.
4. To reform the law.

Direct services to victims attempt to meet the victim's needs for information, emotional support, and advocacy. Many centers have 24-hour telephone "hot-lines" which allow for immediate contact and support after a rape. Information is provided about local hospital and criminal justice procedure, and the victim encouraged to make the necessary decisions about medical treatment, reporting to law enforcement, and communication with family and friends. Counseling services are usually limited to immediate temporary support and short-term follow-up through the use of peer counselors. The goal of counseling is the return of autonomous functioning and control, with prolonged dependency on the center discouraged. Individual counseling is the rule, with group models reportedly less successful. Victim advocacy services include intervention with medical and law enforcement personnel when it is felt that the victim is receiving inappropriate or inadequate treatment, and continuing contact by phone and in person throughout the prosecutory phase. Anonymous reporting is instituted for those victims who choose not to report to law enforcement. Additional services include transportation to hospital, police, and court, as well as babysitters for those times, and a place to sleep either when the victim fears returning home or her home is judged to be unsafe.

Educational programs have as aims self-education, the dissemination of information on rape prevention and resistance as well as what to do if rape occurs, and attitudinal changes. There has been a focus on "demythologizing" the crime of rape particularly with regard to the stereotypic assumptions about both the victim and the assailant. Public education has been accomplished independently, or in coordination with other feminist or social or citizens' groups, or with existing medical and criminal justice institutions.

Educational efforts are viewed as a necessary precursor to law reform as well as to reform of the law enforcement and health care systems. Reform goals have been initiated in cooperation with those institutions, or by assuming a vocal adversary position with regard to institutions. Rape crisis centers have variously recommended or demanded an increase in women police officers, sensitivity to victim needs, and clarification of hospital and police protocol. Many centers have collaborative relationships with both hospital and law enforcement, so that the counselor remains with the victim throughout the required procedures. Center advocates attend rape trials, and on occasion, will pack the court as an additional method of public pressure to change courtroom behavior.

Despite the similarity of goals and services, each program has unique aspects which stem from the nature of the group itself as well as from the resources and attitudes of the community in which the center is located. Some centers remain alienated from professional institutions, while others work closely with existing medical and legal resources. One center reports close collaborative work with the community hospital to the extent that they are called to the emergency room to counsel all victims.[6] Although most counseling is done largely by non-professional women, programs for training these women in the techniques of crisis intervention for rape victims have involved professionals in both medical and legal spheres. Despite the lack of psychological sophistication, counselors have been creative and innovative in providing support to victims and families. It is likely that psychiatric consultative and back-up services would be welcomed by many centers.

Institutional-based rape crisis programs are not considered an improved alternative which replaces the community rape crisis centers. It will be a long time before all victims feel comfortable in reporting directly to hospital and criminal justice authorities. Community groups may tend to attract younger and/or feminist women while more traditional and/or older women will use institutional services. It is anticipated that community programs will continue to be necessary to provide support and to encourage referrals to appropriate medical and law enforcement facilities. Community programs also provide a wide spectrum of services not easily accomplished by the professional, whether it is a safe place to sleep when the victim is fearful of returning home after the rape, or a companion to guide her through the time-consuming and traumatic experiences of the criminal justice procedure should she decide to prosecute. Finally, the community groups continue to provide leadership for educating the public, and this will ultimately be reflected in changes in attitude on the part of the victim as well as the juror.

It will enhance victim treatment for clinicians to familiarize themselves with rape crisis programs and personnel in their communities, with the aim of collaborative rather than competitive services. Many of the larger programs have published guidelines

and informational booklets which will be helpful to both the
clinician and the victim.[6-9] The report of the Center for Women
Policy Studies[5] is recommended as the most comprehensive guide
to citizens' action groups.

References

1. Schmidt, P.: Rape Crisis Centers. *Ms. Magazine,* 14-18, September,
1973.

2. Wasserman, M.: Rape: Breaking the Silence. *The Progressive,* 19-23,
November, 1973.

3. Horos, C. V.: *Rape,* Tobey Publishing Co., New Canaan, Conn., 1974.
(Includes a national directory of rape crisis centers compiled by The Center
for Women Policy Studies, Washington, D.C.)

4. New York Radical Feminists. *Rape: The First Sourcebook for Women,*
ed. by Noreen Connell and Cassandra Wilson, New American Library
(Plume Books), New York, N.Y., 1974.

5. Rape and its Victims: A Report for Citizens, Health Facilities and
Criminal Justice Agencies, *The Response of Citizens' Action Groups: A
Handbook.* Prepared by Center for Women Policy Studies, Washington, D.C.,
1975. Copies available from Law Enforcement Assistance Administration,
Washington, D.C.

6. *Handbook: Medical and Legal Aspects of Rape.* Published by Women
Organized Against Rape, Philadelphia, Pennsylvania, 1973.

7. *How To Start a Rape Crisis Center.* Prepared by the D.C. Rape Crisis
Center, Washington, D.C., 1972.

8. *Stop Rape.* Prepared by Women Against Rape, Detroit, Michigan.

9. Newsletter. Feminist Alliance Against Rape, Box 21033, Washington,
D C. 20009, 1975.

CHAPTER

SIX

REACTIONS TO RAPE

Rape is a personal crisis in the sense that the victim must deal with the impact and meaning of the event for herself. Of at least equal importance, rape is a social crisis in that the significant others in the victim's life will also be strongly affected by the event, with the possibility of negative consequences in the relationship between the victim and husband or boyfriend, or in the case of child victims, the parents and schoolmates.

Some knowledge of crisis theory is basic to one's understanding of the crisis which rape precipitates. Gross stress has been defined as any unusual influence or force perceived as a threat to a vital goal or need of an individual. Stress reactions represent attempts to defend or restitute the personality from disorganization. While reactions will vary according to one's prior adaptive capacity, coping style, and social support, there seem to be four clinical phases of response,[1,2] even in adequate life adjustments.

1. Anticipatory or threat phase in which there is a fine balance struck between the need to protect one's illusion of invulnerability "that could never happen to me" while maintaining sufficient reality perception to prepare and protect oneself from real danger. Where a stress is planned, as in elective surgery, working through prior to the event protects ego integrity.

2. Impact phase which lasts for the duration of the stress. Variable degrees of personality disintegration occur which include physiologic patterns of anxiety as well as heightened alertness and attention focused on the present. If the stress is overwhelming, this state of increased vigilance may be followed by diminished alertness, numbness, dulled sensory, affective and memory functions, and disorganized thought content. Thus, in one sample,[3] 12-25% of victims were "cool and collected"; 10-25% exhibited inappro-

priate responses with paralyzing anxiety, hysteria, confusion and loss of control; and the majority were stunned and bewildered.

3. Recoil phase in which there is beginning return of emotional expression, self-awareness, memory, and behavioral control, in the context of a limited perspective and heightened dependency needs. The individual becomes aware of his/her adaptive and/or maladaptive responses to the stress. There is the potential for either increased self-confidence or damage to one's self-esteem, depending on positive or negative perceptions of one's behavior during the stress. During this phase of personality reorganization, support or nonsupport by others becomes a critical issue for subsequent psychological events.

4. Post-traumatic phase begins when one's sense of self has been maximally reconstituted. A successful response includes recall of the stress with assimilation and affective sharing of the experience, and spontaneous recovery. Maladaptive responses may result in a permanently impaired concept of self and world with evidence of continuing anxiety and depression, anger, guilt, nightmares, and an impaired functional capacity.

Prevention of maladaptive stress reactions may be facilitated by adequately supporting the anticipatory phase and diminishing the intensity of the abrupt onset in the impact phase. Neither of these measures is possible in the event of rape. Bard and Ellison[4] present an excellent summary of those factors which result in crisis as well as a practical model for conceptualizing the specific difficulties encountered by a crime victim in crisis. They define a crisis as a subjective reaction to a stressful life experience, one so affecting the stability of the individual that the ability to cope or function becomes seriously compromised. Characteristics of stressful situations which result in crisis reactions include suddenness so that there is no preparation time, arbitrariness in that the event is perceived as unfair and capricious, and unpredictability which stands in contrast to the normal developmental or anticipated life crisis. Although crisis reactions differ according to one's adaptive capacities, there are a set of fairly predictable responses, among them 1) disruption of normal patterns of adaptation with disturbances in eating and sleeping, diminished function, attention, and concentration span 2) regression to a more helpless and dependent state in which support and nurturance are sought outside of the self and 3) increased openness and accessibility to outside intervention, which present a unique opportunity to affect the long-term outcome of the crisis. It is important to recognize that the disruption associated with a crisis may appear acutely, or may come as a delayed reaction.

It was only six years ago that the first clinical study of rape victims was reported[5] by Sutherland and Scherl. The authors found similar sequences of events among the 13 victims interviewed. The women were young social workers all of whom became victims of circumstances which they did not precipitate and for which they could not be held responsible. Phase one, or the acute reaction, was characterized by signs of acute distress which included shock, disbelief, emotional breakdown and disruption of normal patterns of behavior and function. She is unable to talk about what has happened, and is uncertain about telling significant others, much less reporting to the hospital or police. Guilt may be prominent, with fears that her own poor judgment precipitated the event. During this phase of overt disruption, there are many concrete concerns which demand fairly immediate attention: Should she tell her family, husband, lover, friends, children what has happened? What are the implications of not telling significant others? Will there be newspaper publicity? Will friends and neighbors find out? What is the likelihood of pregnancy or venereal disease? Should she report to law enforcement? Should she prosecute? Will she be able to identify the rapist? Is she in danger of another assault by the assailant? Phase two, that of outward adjustment, is ushered in within several days to weeks after the rape with temporary resolution of the immediate anxiety provoking issues. In an attempt to overcome her anxiety, the victim returns to her normal life patterns and announces that all is well, reassuring both herself and those close to her. Sutherland and Scherl comment that "this period of pseudoadjustment does not represent a final resolution of the traumatic event and the feelings it has aroused. Instead, it seems to contain a heavy measure of denial or suppression. The personal impact of what has happened is ignored in the interest of protecting self and others." Phase three, that of integration and resolution, often goes unrecognized. Depression is prominent along with the need to talk about what has happened. The two major themes which emerge include the need to integrate the event with her view of herself and to resolve her feelings about the assailant. "Her earlier attitude of 'understanding the man's problems' gives way to anger toward him for having 'used her' and anger toward herself for in some way having permitted or tolerated this 'use'."[5]

Burgess and Holmstrom[6,7,8] have done the most comprehensive clinical analyses to date on the crisis associated with the stress of rape. Their study population consisted of all rape victims seen in the emergency ward of Boston City Hospital over the one year period 1972-1973, with a wide spectrum of social class, life style, age, and work status represented in their sample. They define a specific rape trauma syndrome as a two stage process: the immediate or acute response in which the victim's life style is completely disrupted by the rape, and a long-term process in which the victim must reorganize her disrupted life style. The syndrome includes the physical, emotional and behavioral reactions which occur as a result of the encounter with a life-threatening event.

Immediately, one sees a wide range of acute impact reactions. Emotional responses fall into two major categories, an expressed style and a controlled style. The expressed style is manifested by visible evidence of acute anxiety, fearfulness, sobbing, shaking, or other indications of a massive affective response. The controlled victim appears calm and collected with little external evidence of distress, and it is often commented upon by the victim herself who seems aware "that this will hit me later." The apparent calmness may reflect shock and disbelief, as well as exhaustion, since rapes usually occur at night, so that the victim likely has not slept in a 24 hour period. Burgess and Holmstrom comment that while the expected response to rape is that of shame and guilt, this was not supported by their analysis of acute responses. The primary reaction acutely is related to the fear of physical injury, mutilation and death, that is, the awareness that they could have been murdered. Mood swings are common and include feelings of humiliation, degradation, guilt, shame, embarrassment, self-blame, anger, revenge, and fear of another assault by the assailant. The primary defense is to block the thoughts from her mind, although they continue to haunt her. The wish to undo the event is reflected in fantasies of how she might have handled the situation differently, thereby avoiding the assault.

The wide variety of physical and physiological reactions occurring during this time period depend in part on the location and extent of the injury. Complaints may refer to general feelings of soreness or to more localized chest, throat, arm or leg pain. Specific symptoms may occur which relate to the area of the body which has been assaulted. Victims of oral rape may have a variety of mouth and throat complaints, while victims of vaginal or anal penetration have different physical concomitants. Sleep disturbances are the rule, with insomnia and wakefulness. Victims who were attacked in bed may wake up in terror at the time that the rape occurred. Appetite disturbances include anorexia, nausea, and vomiting, which must be differentiated from similar symptoms caused by hormones administered to prevent pregnancy.

This acute phase may last from a few days to weeks with a gradual merging into the long-term reorganization process. The manifestations of this second stage are quite variable, and will depend on the victim's personality style, available support system, and the treatment she encounters from others. One might expect a lengthening of this process, for example, for the victim who decides to prosecute, the criminal justice procedures conceivably being drawn out over a one or two year period. The following constellation of symptoms comprise the usual pattern for this phase.[8] Changes in life style are prominent, with impaired level of function at work, home and school. Many women feel the need to get away and actually move to another residence or another city, while others are fearful of leaving their homes at all, or give up autonomy by returning to their families. Sleep disruptions continue with vivid dreams and nightmares. Dream content reflects concerns about violence and falls into two categories, reflecting either a thinly disguised re-enactment of the actual rape or mastery of the rape in which the victim commits a violent act. Phobias appear which seem specifically determined by the nature of the rape experience, with fears of crowds or of being at home or outside depending on the location of the assault. An inadvertent physical contact on the street or the appearance of someone resembling the assailant may precipitate a panic reaction. Sexual fears are common, with a decline of interest as well as withdrawal from one's partner.

If this represents the normal reaction to rape then one can assume a more intense and complex reaction will occur in certain circumstances. The victim who is raped by a trusted friend or relative is faced with a greater burden of resolution than would be the case if the assailant is a stranger. The assailant treats the victim as an object, and if she also perceives herself as an object, she can isolate and depersonify the event. This is not possible with an intimate friend. The existence of psychological problems and/or maladaptive behavior patterns prior to the rape increases the likelihood of maladaptive coping strategies following the rape. Serious medical problems either prior to or as the result of the rape may affect the outcome. In one such case, a middle-aged woman with chronic renal insufficiency was raped. Subsequently, depression was reflected in her affect, appearance of gastrointestinal symptoms, and refusal to take the medications necessary to maintain her already precarious chemical balance. She died two months following the rape. Other coincident life crises involving the victim's family, social network, academic or work status may compound the stress. Indeed, there is some preliminary evidence[9] that a major personal crisis increases the possibility of rape occurring, in that one's energy is focused on the crisis at hand, with insufficient attention to whether or not one places oneself in a vulnerable situation. The rape, then, may be followed by disabling depression with or without suicide attempts, alcoholism, drug abuse, frank psychosis, or a marked increase in physical symptomatology, as with one victim who refused counseling services, but sought multiple medical evaluations for severe headache.

Burgess and Holmstrom[7,8] describe a "silent reaction" which may alert the clinician to an earlier and unresolved rape occurring months to years earlier. In their sample the following cluster might suggest a previous rape experience: increased anxiety as the interview progresses, history of a sudden increase in irritability with avoidance of men and sexuality, reports of an acute onset of phobias and violent nightmares, and a loss of self esteem. Psychiatrists in their psychotherapeutic work not infrequently uncover a rape occurring in the past, the therapist often being the first person ever to be told this "secret." Most rape counseling programs have seen victims for whom this is not the first sexual assault, and it is suspected that there are some areas in which rape is endemic as well as some women who are especially vulnerable to rape.

In addition to the stresses already enumerated, there are also age-specific issues for the victim which vary with the developmental phase. Notman and Nadelson[2] as well as Burgess and Holmstrom have described these life stage considerations. The single young woman is more likely to be sexually inexperienced, with encounters with men limited to the trusting, caring figures of her childhood, or high school dates. If the rape is her first sexual experience, she may be quite confused about the relationship between sexuality, violence, and humiliation. If she is engaged in separation from family and the establishment of an independent identity, her sense of adequacy is challenged by her feeling that she really cannot take care of herself. A college student living away from home may decide not to tell her parents because of their possible insistence that she leave school and return home. The integrity of her body is also at issue, and is reflected in concerns about the pelvic examination which may be a new experience and perceived as like the rape itself. The woman who has an ongoing sexual relationship may choose not to tell the man that the rape has occurred for fear of disrupting the relationship. There is considerable reality to these fears, so that her silence protects the relationship but leaves her feeling guilty and without support. The divorced or separated woman is in a difficult position because her life style, morality and character are even more frequently questioned. The rape experience may confirm her fears of inadequacy about independence and autonomy. The woman with children must deal with the problem of what, how and when to tell them. Since the event may be known in the community, this too must be managed with the children. The middle-aged or older woman will have concerns about independence. She may already be in a crisis phase about her life role, with changes in her relationship to husband or children, role loss, and absent family members. The misconception that an older woman has less to lose by a rape than does a younger woman may seriously impede her ability to resolve the rape crisis.

References

1. Weiss, R. J., Payson, H. E.: Gross Stress Reaction I, in A. M. Freedman and H. I. Kaplan (eds.). *Comprehensive Textbook of Psychiatry*, Baltimore, Williams and Wilkins, 1027-1031, 1967.

2. Notman, M. T., Nadelson, C. C.: The Rape Victim: Psychodynamic Considerations. *American Journal of Psychiatry* 133:4, 408-413, April 1976.

3. Tyhurst, J. S.: Individual Reactions to Community Disaster: The Habitual History of Psychiatric Phenomena. *American Journal of Psychiatry* 107, 764-769, 1951.

3. Tyhurst, J. S.: Individual Reactions to Community Diseaster: The Habitual History of Psychiatric Phenomena. *American Journal of Psychiatry*, 107, 764-769, 1951.

4. Bard, M., Ellison, K.: Crisis Intervention and Investigation of Forcible Rape. *The Police Chief*, 41:5, 68-73, May 1974

5. Sutherland, S., Scherl, D.: Patterns of Response Among Victims of Rape. *American Journal of Orthopsychiatry*, 40:3, 503-511, April 1970.

6. Burgess, A. W., Holmstrom, L. L.: The Rape Victim in the Emergency Ward. *American Journal of Nursing*, 73:10, 1740-45, October 1973.

7. Burgess, A. W., Holmstrom, L. L.: Rape Trauma Syndrome. *American Journal of Psychiatry*, 131:9, 981-86, September 1974.

8. Burgess, A. W., Holmstrom, L. L.: *Rape: Victims of Crisis*. Bowie, Maryland, Robert J. Brady Co., 1974.

9. Nadelson, C., Hilberman, E.: personal communication.

CHAPTER

SEVEN

COUNSELING AND TREATMENT CONSIDERATIONS

It should be apparent from the elaboration of the rape trauma syndrome that it is not possible to provide a simplistic approach to the treatment of the rape victim and her family. All recent studies stress the need for immediate availability of crisis counseling for victims, with some indications that continued counseling, either formal or informal, may be necessary for a period of three to twelve months. There seem to be three assumptions regarding counseling about which there is a consensus: 1) Crisis intervention will facilitate working through of the trauma and diminish the likelihood of long-term psychopathological consequences; 2) The victim needs emotional support from whomever she comes in contact in the crisis period; 3) Rape is a crisis for significant other family members and friends who also need emotional support.

These few assumptions, however, may be the only points of agreement in the area of emotional support systems. The report of the Center for Women Policy Studies[1] summarizes the debate over peer counseling versus professional mental health services as follows:

> In the instances of drug abuse as well as rape victim counseling, when nonprofessional people began to take specific mental health and societal concerns into their own hands, considerable alarm was registered among mental health professionals. However, many professionals have come to agree that they have developed no special theory of treatment best suited to rape victims, and that it is in the best interest of their profession to participate in what is becoming a rapidly spreading counseling movement.
>
> Although there is consensus that immediate crisis intervention counseling is needed, there is confusion and ambivalence among both professionals and paraprofessionals about the degree of emotional support and counseling required to facilitate a victim's adjustment or prevent the development of psychological or behavioral problems resulting from the rape. The Medical Society of the District of

Columbia in a report of its Ad Hoc Committee on Sexual Assault Victims concluded that the trauma of rape is so serious that professional psychological help should be immediately available. On the other hand, feminists tend to believe that it is detrimental to suggest to a victim that she needs professional help unless there is some indication she cannot cope with the reality of the assault or its consequences. They believe most victims simply need a sympathetic and accepting listener and enough information to enable them to make choices.

Rape crisis center counselors usually believe they can tell when a woman needs help that they are not prepared to give. Some have professional volunteers they can call for advice, and most have lists of philosophically and personally acceptable professionals to whom they can refer cases. In Salem, Oregon, professionals remain on call in their offices one hour per evening for consultation and emergency help. Many psychologists and social workers volunteer to train and supervise lay counselors and become active, valuable, and accepted personnel of the rape crisis center. The center in Cambridge reported that the utilization of professionals in this way is an excellent educational technique and stabilizing influence for the counselors.

There remains, however, sharp resistance to the use of salaried mental health professionals as rape crisis intervention counselors. It stems in part from the belief of rape crisis centers that professionals as a group are not more informed about the psychological aspects of rape and, in fact, have learned much of what they know from the rape crisis movement. In addition, the women's movement has a generally skeptical attitude toward most psychiatric treatment, which they view as reinforcing the oppression of women in traditional roles, including that of the seductress. Finally, rape crisis centers that are attempting to find financial support for their programs resent having to compete with professionals who would not have been interested in the problems of rape victims had they not been educated by the rape crisis movement.

Notman and Nadelson[2] also comment on the attitudes of professionals regarding rape victims, professionals sharing the societal myth of the victim as a young, attractive and seductive woman who could have avoided the assault if she really wanted to.

Until recently many psychiatrists have felt that rape was not a psychiatric issue and further that psychiatrists had little to offer the rape victim. They often shared the prejudiced view that the victim "asked for it." The victim may be seen as acting out her unconscious fantasies and therefore not "really" a victim. Thus the rape victim has not been offered the sympathetic understanding usually extended to people in crisis. Rape falls into a group of emotionally charged issues, where prejudice prevents the objectivity which would be available if it were regarded as a traumatic event.

It is not the function of the clinician to decide whether the victim has "really" been raped. Rape is a legal and not a medical term. The fact that the victim perceives herself as having been violated remains the significant event.

Throughout this chapter, then, the deliberately vague terms "counselor" and "counseling" are used to reflect the fact that a variety of professionals and nonprofessionals will have occasion to take part in the victim's care. Counseling services have been provided by specially trained nurses, and by nonprofessional women trained as part of a community rape crisis program, as well as by mental health professionals.[1-6]

In the immediate aftermath of the rape, issues of personal safety and control emerge as primary concerns. The victim has just had an experience in which her very existence has been threatened, with total loss of control of what was done to her. Her immediate needs are for a sense of physical safety as well as for assistance in assuming some control over what has happened to her and what will happen to her in her dealings with both individuals and institutions. The presence of an empathic and supportive individual, who may be a counselor, clinician, or friend, will enhance her sense of safety. In no event should she be left alone. She must be allowed to talk and to ventilate her feelings, with reassurance and validation of her responses. Whether this occurs in the police station, the physician's office, or the hospital emergency room, the issues are the same. She is dealing with new personnel, institutions and procedures, and it becomes the responsibility of the designated counselor to help her regain control by preparing her for subsequent events. Thus, the counselor must have available information about medical and legal procedures, so that the victim can make choices based on informed consent. The victim may or may not wish to report to the hospital or law enforcement, and this is her right. While the counselor may have some opinions about what she ought to do, coercion obviously has no place in the management of the rape victim.

Burgess and Holmstrom[4] summarize the assumptions they make in counseling victims. Their context is a short-term issue-oriented crisis model, with the focus on the rape incident, and the goal of restoring the victim to her previous level of function as quickly as possible. Since rape represents a crisis which disrupts the victim's life style in four areas, physical, emotional, social and sexual, all of these issues are appropriate areas of scrutiny. The victim is assumed to be normal, that is, an individual who was managing her life adequately prior to the crisis. The rape is viewed as a crisis situation and previous problems unassociated with the rape are not considered priority issues for discussion. When other issues of concern are identified which would require the use of a different model, referrals are offered to the victim. In contrast to the traditional model in which the patient is expected to be the initiator, the counselor takes a more active role in initiating follow-up contact. Burgess and Holmstrom report excellent results with telephone counseling although other clinicians have had less success with this mode.

An analysis of rape crisis requests[4] suggests that there are four major categories of need, namely for police intervention, medical intervention, psychological intervention, and control (especially with victims who present in an incoherent state). There is an additional group who may not perceive that anyone can be of help to them, with ambivalence if not hostility about any intervention.

The following summarizes appropriate areas for counseling intervention in the immediate phase:

The assault: the circumstances; her relationship to the assailant; conversation with the assailant; sexual details; physical and verbal threats; amount and kind of resistance; the use of alcohol or drugs by either victim or assailant; the victim's emotional and sexual reactions. The victim will certainly have feelings about each of these issues, and the counselor's ability to gently unravel the details of the assault will clarify which of these concerns are most problematic. The focus may be on the rape as a first sexual encounter, or intense guilt for not putting up a fiercer struggle. Brussel[7] notes that "the victim who is beaten into senselessness may suffer far less emotional trauma than the woman who submits to rape when her life is threatened." The victim's coping strategy before and during the assault will become an issue after the assault.[8] Victims who are raped in the context of drinking, hitchhiking or a date are likely to hold themselves accountable.

Medical concerns: the quality of her treatment in the hospital; interactions with medical and nursing staff; extent and type of physical injury; the possibilities of pregnancy and venereal disease; the pelvic examination. The pelvic examination is invariably affect-laden, and a frequent focus for the victim's anger ("It's like being raped again") so that adequate preparation is important. Victims also express many fears about their bodies being irrevocably changed and this too deserves close scrutiny.

Law enforcement: does she plan to report; if she has reported, how was she treated; what kinds of questions was she asked; was she encouraged or dissuaded from considering legal charges against the assailant. She may have reason to feel intensely frustrated and angry with police and/or hospital and part of the counseling process is to help the victim vent her feelings about her treatment, as well as to validate her feelings.

Prosecution: does she plan to press charges; cooperate with the police investigation; appear in court; and how does she feel about this? This is often an issue of high priority for the victim, who must make this decision at a time when it is difficult to think rationally about anything. Concrete information is usually necessary because most people are naive about the criminal justice system. Some victims think they are breaking the law by not reporting and prose-cuting, and this may need clarification. In contrast, a victim may herself have been charged or convicted of a violation in the past and this too will affect her attitudes. The victim's rationale for not wishing to prosecute may be quite realistic given the treatment of victims by the courts, or may reflect her fears that she did indeed precipitate the rape. The counselor's role is to help the victim clarify for herself how she feels about this, and to support the victim's decision, whatever it may be.

Social support system: what people constitute her family; and friends; whom does she plan to tell about the rape; what responses does she anticipate? How the rape may affect or change her rela-tionship with significant others is invariably an issue in the immedi-ate crisis period. An evaluation of her support system will determine in part the extent to which continued counseling becomes necessary, that is, the availability of a strong social support network will diminish the need for ongoing counseling intervention. In contrast to most personal crises where the counselor or therapist reinforces the importance of sharing the crisis with significant others, no firm guidelines exist about communicating the event of the rape because of the real possibility that the revelation may disrupt the relationship. Thus, a husband may perceive the rape as an extramarital affair, and the parents of an adolescent may project their own sense of guilt with such remarks as "I told you something like this would happen . . . if you had only listened to us." Each victim then must be seen as a separate individual regarding the issue of whom to tell. Where the victim does share the information with significant others, the counselor has an important role in family counseling. The victim's social network must be sensitized to the meaning of the rape so that they are able to give honest support to the victim. Where there are preexisting conflicts in a relationship to a significant other, the rape may be invoked as another weapon in an ongoing battle. The counselor's intervention with family and friends is as important as the intervention with the victim. In one case, the counselor supported the victim for confronting her boyfriend in the emergency room with the evidence of his disbelief, and this was thought to be a significant factor in the ability of both to work through the crisis with mutual support. In cases where the victim chooses not to tell close friends or family members, guilt or distancing may occur, for which the victim will need counseling assistance.

Physical safety: If she lives alone, or was raped at home, she will likely be fearful of returning home. Further, the assailant (who has usually not been apprehended) may have threatened to return, and the victim is fearful of this even when the threat has not been made. A temporary resolution may involve identifying a friend with whom she can stay as well as available options for whom to contact in the future.

Preparation for victim responses: As previously described, there is a wide spectrum of immediate responses to rape. A victim may not wish to talk and may show little if any affect. In any case, it is important for the counselor to prepare the victim for the range of reactions likely to occur, so that the victim becomes sensitized to her own feelings, and has some awareness that the inevitable disruptions in lifestyle are normal. Mild sedation for sleep is usually indicated as well. Knowledge that the victim can call upon the counselor throughout this period provides additional reassurance that she will not be left in isolation.

Although the counselor's role with the victim in the later re-organization phase will not be specified here because of the obvious complexity of victim needs and responses, some general remarks about the dynamics of the victim's response to rape seem appropriate, particular in regard to affect and unconscious fantasies. Notman and Nadelson[2] point out that the universality of rape fantasies hardly makes every woman a willing victim and every man a rapist. Fantasies usually do not picture the actual violence of the experience, and while all individuals have aggressive if not homicidal fantasies, most people keep thought and action separate. The victim has had an overwhelming confrontation with the assailant's loss of control of sadistic and aggressive impulses, which serves to challenge the woman's confidence in her own defenses and controls. Further, women usually relate ambivalently to men, perceiving them as both protectors and potential aggressors. The confrontation with this violent potential has as a result a marked decrease in the woman's ability to trust men, as well as diminished trust in her own capacity to control impulses.

There is surprisingly little clinical evidence of rage in rape victims, compared with the outrage stimulated in others hearing of the rape. The anger, when it does occur, is likely to be a delayed experience. The absence of anger may be the result of multiple etiologies, involving both sex-role socialization norms in which women are discouraged from showing outward aggression as well as environmental reinforcement for the idea that rape is a justifiable retaliatory act. One victim described the impact of rape by stating "I have become a very passive person" while another sought psychiatric help two years following rape with the chief complaint "Help me to get angry." The universality of guilt and shame stand in striking contrast to the absence of anger. The victim mirrors the courtroom trial with her own concerns about her unconscious fantasies, her behavior in the actual rape situation, and a focus on the sexual aspects of the experience. Issues of morality, virginity, and fidelity enhance her sense of herself as bad or dirty.

Little information is available about the long-term consequences of rape because our awareness that rape is a crisis is of recent origin. Borges[9] makes some preliminary comments on a study she is conducting on female suicide attempters, in which there seems to be a relationship between an earlier rape experience and suicide attempts. Clinical evidence of the silent rape reaction is witnessed by psychiatrists who uncover an earlier rape, and find that the event is still very much alive and unresolved. While it is difficult to predict all of the long-term needs of the victim, some of the issues which seem to reemerge at a later date are mistrust and avoidance of men, sexual disturbances, phobic reactions, and anxiety and depression often precipitated by seemingly trivial events which symbolize the original trauma. With the shift to a more open attitude about rape, systematic study of the long-term consequences and treatment implications will finally become possible.

References

1. Rape and its Victims: A Report for Citizens, Health Facilities and Criminal Justice Agencies, *The Response of Citizens' Action Groups: A Handbook*, prepared by Center for Women Policy Studies, Washington, D.C., 1975. Copies available from Law Enforcement Assistance Administration, Washington, D. C.

2. Notman, M. T., Nadelson, C. C.: The Rape Victim: Psychodynamic Considerations. Presented at Special Session on Rape, APA, Anaheim, California, 1975. Publication pending, American Journal of Psychiatry, April 1976.

3. Burgess, A. W., Holmstrom, L. L.: The Rape Victim in the Emergency Ward. *American Journal of Nursing*, 73:10, 1741-45, 1973.

4. Burgess, A. W., Holmstrom, L. L.: *Rape: Victims of Crisis*. Bowie, Maryland, Robert J. Brady Co., 1974.

5. Hayman, C., Lanza, C.: Sexual Assault on Women and Girls. *Journal of Obstetrics and Gynecology*, 109:3, 480-486, 1971.

6. Wasserman, M.: Rape: Breaking the Silence. *The Progressive*, 19-23, November 1973.

7. Brussel, J.: Comment Following Menachim Amir, Forcible Rape, *Sexual Behavior*, 1 (8).

8. Burgess, A. W., Holmstrom, L. L.: Coping Behavior of the Rape Victim. Presented at Special Session on Rape, APA, Anaheim, California, 1975. Publication pending, American Journal of Psychiatry, April 1976.

9. Weiss, K., Borges, S.: Victimology and Rape: The Case of the Legitimate Victim. *Issues in Criminology*, 8:2, 71-115, Fall 1973.

CHAPTER

EIGHT

THE CHILD RAPE VICTIM

In preparing this monograph, it was anticipated that there might be a larger body of information about the child victim because social attitudes about the sexual molestation of children would dictate more empathy than is the case for adult victims. Although child abuse and incest have become relatively popular subjects for scrutiny, the literature about children who are raped is scarce. Further, the attitudes reflected therein are for the most part identical to those which prevail about all rape victims. In a recent brief guide to office counseling in a case of sexual molestation of a child,[1] the opening statement is "The child often plays some part in encouraging the sexual situation." The author goes on to state that "repeated sexual involvement with the same person says clearly that at some level the child wanted the relationship to continue" and finally and remarkably, he urges the clinician to "use a supportive, nonjudgmental approach" to dealing with the situation.

The few articles in the literature either focus on the medical aspects of treatment[2-4] or criminal justice implications.[5-7] There is superficial attention given to the emotional trauma of the event, other than instructing the clinician to have the pediatrician or family doctor see the victim in follow-up or to "refer the patient to a child psychologist or psychiatrist for the prevention or treatment of any resulting psychoneuroses."[2] The one article on immediate management of a child victim written by child psychiatrists[8] reports a series of nine cases, the authors stressing the point that only one of these was a "true rape." These authors cite the case of a 12 year old girl who "was grabbed by two men on the street and dragged into an abandoned home. A witness confirmed this story. Her unconscious complicity is suggested by her stated wish in the emergency room to bear the baby if she were pregnant and to keep it." The same authors, however, do acknowledge that while a seduction may have different qualities than an attack, even a seductive child is not expected to have full adult comprehension of the act she is courting, and is therefore not responsible. Nonetheless, the focus of attention is on the complicity of the victim rather than the trauma of the event.

Burgess and Holmstrom[9] present the only description in the literature of the manifestations of rape trauma in children and adolescents. In general, the rape trauma syndrome is similar to that experienced by adults. The differences are consistent with the developmental stage of the young victim, and the alternate affective and behavioral modes by which a youngster responds to stress. Thus, nausea, vomiting, and bedwetting may occur along with the spectrum of life-pattern disruption seen in adults. In addition to reacting to the threat to her life, feelings of embarrassment were prominent, especially in adolescent victims. Many of the concerns of children focused on how the event would affect them at school and the anticipated reactions of their peer group. Of the 23 victims described, all but three experienced mild to severe symptoms during the long-term reorganization phase, with nightmares, phobias of being left alone, and panic reactions on seeing the assailant or the crime scene or a symbolic reminder of the assault. Sexual fears were prominent, and were difficult for the victim to discuss, since the issue was likely to be new, unfamiliar, and embarrassing. Problems at school were evident, and the wish to change schools or the development of a school phobia seemed to be the child's equivalent of the adult's need to move to a new residence or town.

Families also respond to the rape with affective responses which may be as intense and disruptive as they are for the victim. There may be surprising inability of parents to provide adequate emotional support to the child, even when they have been supportive in other life crises. The mother may be struggling with denial of her own anger for having been raped literally or figuratively by adverse life circumstances, while both parents may feel guilty for not having been able to protect the child. The need to blame someone is prevalent in the acute stage, and there are three targets for this: the assailant, the child, and themselves. Parents express anger for the child's "stupid" behavior and may punish the child by restricting privileges. Where the victim is an adolescent, parental concerns about her sexual provocation may surface. When parents assume the full responsibility for the event themselves, they feel that they have been inadequate parents, or shouldn't have allowed the child to go out, or shouldn't go out themselves. In some circumstances where a relative or baby-sitter molests a child, parents may have previously disbelieved the child until there is gross evidence to support the youngster's complaint, thus leaving the parents devastated by guilt. The more long-term issues which surface are those related to sexuality, which may be difficult for the parents to handle with the child especially if there has been no prior discussion about sex.

Compound reactions occur when the child has a past or current history of physical, psychiatric or social difficulties, with running away from home one of the reported behaviors. As with the adult, the child may have a silent reaction in which she tells no one about the rape. Burgess and Holmstrom suggest that the following symptoms in children may serve to reflect that an assault occurred: acute onset of somatic symptoms, gastrointestinal symptoms, sleep disturbances and enuresis in a child under ten; and marked withdrawal from one's usual activities and relationships at home and school.

While most of the literature on children focuses on whether or not a "real" rape occurred, it is likely that in many more cases, the child will not report a rape. In one case a teenager refused the medical examination insisting that she was assaulted but "got away" without being raped. Subsequently, it appeared that she was fearful that her parents would blame her, and found it easier to deny the actual rape to them. In those cases where a complaint of rape appears not to be substantiated, one can still assume that the child and/or family is in a crisis which demands the attention of a mental health professional.[10] Parents who bring a child to the hospital with a history of rape may be wanting confirmation of the child's chaste status, just as a child with a rape complaint may see no other way to raise sexual issues, or to gain information about pregnancy prevention or venereal disease. Where parents need to deny sexuality, the child may be forced to pretend there was a rape rather than consensual sexual intercourse. In any event, there is a clear indication for psychiatric intervention in such instances.

The strongest and most important support system the child has is her family. The focus of counseling must then be to facilitate communication and empathy among family members both in terms of their own needs as individuals in crisis and their ability to support the victim through the crisis phase. There will be wide variations in parents' perceptiveness and willingness and ability to talk openly about the rape, and having someone available with whom to discuss the difficulties of such communication is helpful. The child needs encouragement to return to her usual life-style as quickly as possible, and the counselor may need to be involved with the parents, the child, and at times, the school. The counselor must sensitize parents to symptoms of distress in the child, so that the parents are aware of which kinds of responses are usual and which ones herald more complex problems. How one talks to a child about the details of the rape and her feelings about it depends on the age of the child, but one can usually assume that children know and remember a great deal more than adults suspect, so that there is no justification for not speaking directly with the child.

The following case report illustrates the conspiracy of silence which often follows the rape of a child, as well as the dilemma posed for the victim and her peers at school.[11]

Mary is a first-grader who was raped on her way home from school. She showed obvious signs of distress on her return to school, with intermittent anxiety and depression, as well as isolation from peers. As part of an experimental program, her class met weekly to discuss feelings about growing up, becoming a first-grader, and leaving the old relationships within the home. The teacher had successfully led the group through complex and highly charged topics, such as trouble between students, and how severe problems at home affected school work. At the time of the first group meeting after the rape, the teacher accurately described Mary's affective and behavioral responses to the psychiatric consultant, but felt extremely reluctant to discuss rape with the children. The consultant agreed that it was a new and frightening topic for the group. He noted, however, that Mary's behavior was evident to all of her peers, and that every child had probably heard some version of what had occurred, likely with considerable anxiety.

The teacher allowed the consultant to initiate a discussion with the children, beginning with "what has happened to Mary?" It became rapidly apparent that every child did indeed know of the rape, and that this was a terrible event. Most, but not all, knew what rape meant. During the discussion, the children were able to verbalize their fears that something like that could also happen to them. They also expressed their uncertainty and discomfort about how to act towards Mary when she was acting so different from her usual self. After this relatively simple level of open talk, it required only a mild suggestion from the consultant that perhaps Mary would like her friends to treat her as they usually did.

The reality is that rape could indeed happen to any child in the class. This awareness was unchanged, and was already known by the children. What did change was the secondary peer isolation which stems from their need to pretend that the event didn't happen, as well as fear of their inability to deal with the victim's feelings. This kind of denial reverberates throughout the victim's social network, and is probably intensified when the victim is a child.

Finally, the decision of whether or not to prosecute is particularly complex where a child is involved, and there must be close attention to the additional stress imposed on the child and family before this decision is made. De Francis[12] reported that more than one thousand court appearances were required for the prosecution of 173 cases of sexual abuse of children. The special problems of child victims in court have not been adequately dealt with by mental health professionals or the criminal justice system. The child victim rarely goes into court with any advance preparation other than some sporadic attempts at rehearsing testimony. Attorneys have not been sensitive to the impact and many meanings testifying can have on a young victim. The mental health professional may be too absorbed in the impact of the assault on the child to pay much attention to the problems which surround the court appearance. The child must present the facts of the assault in an open courtroom setting and is subject to cross examination by defense. The courtroom setting is imposing to mature, self-confident adults. To young victims, it is shattering. Their shame and guilt is compounded by the voyeuristic aspects of open courtroom testimony. In addition, those who question the child victim are not usually sensitive or trained to specialized approaches with children. The trauma to a child of such an experience may be equivalent to or greater than the trauma of rape. Unfortunately, because of the general reluctance of our society to deal with the problems of a rape victim, the highly complicated problems of the child rape victim and her relationship to the courts have only recently been described.[5-7] Attempts are being made to streamline legal procedure where children are involved, and the decision to prosecute will depend on the responses of the child and family as well as the criminal justice procedure and personnel in a given community. In the event that the family decides to proceed with prosecution, it is likely that prolonged counseling support will be necessary.

The long-term impact of rape on a child has yet to be studied, with Anny Katan's report being one of the few in the literature.[13] She describes her analytic work with six adult women who were raped in childhood. The common theme for all six women was "unbelievably low self-esteem" which Katan attributes to a severe disturbance in the fusion of sexual and aggressive drives:

> My patients had no criminal or delinquent tendencies. Their aggression was turned against the self in a savage form . . . they were fragmented, they could never feel that they were whole persons. They regarded their own aggression as dangerous and raw, whether it was turned against the self or against the outside world. They felt keenly that the traumas had caused irreparable damage.

Katan's powerful description provides added confirmation that even the child victim understands the violence which defines the rape encounter.

References

1. Kiefer, C. R.: Sexual Molestation of a Child. *Medical Aspects of Human Sexuality*, 127-28, December 1973.

2. Breem, J. L., Greenwald, E., Gregori, C. A.: The Molested Young Female: Evaluation and Therapy of Alleged Rape. *Pediatric Clinics of North America*, 19:8, 717-725, 1972.

3. Robinson, H. A., Sherrod, D. B., Malcarney, C. N.: Review of Child Molestation and Alleged Rape Cases. *American Journal of Obstetrics and Gynecology*, 110, 405-6, June 1971.

4. Capraro, V. J.: Sexual Assault of Female Children. *Annals of New York Academy of Science*, 142, 817-19, 1967.

5. Flammang, C. J.: Interviewing Child Victims of Sex Offenders. Chapter 14, *Rape Victimology*, ed. LeRoy G. Schultz, Charles C. Thomas, Springfield, Illinois, 1975.

6. Schultz, L. G.: The Child as a Sex Victim: Socio-Legal Perspectives. Chapter 15, *Rape Victimology*, ed. LeRoy G. Schultz, Charles C. Thomas, Springfield, Illinois, 1975.

7. Libai, D.: The Protection of the Child Victim of a Sexual Offense in the Criminal Justice System. Chapter 17, *Rape Victimology*, ed. LeRoy G. Schultz, Charles C. Thomas, Springfield, Illinois, 1975 (originally published in *Wayne Law Review*, 15, 1969).

8. Lipton, G. L., Roth, E. I.: Rape: A Complex Management Problem in the Pediatric Emergency Room. *Journal of Pediatrics*, 75:5, 859-866, 1969.

9. Burgess, A. W., Holmstrom, L. L.: *Rape: Victims of Crisis*. Bowie, Maryland, Robert J. Brady Co., 1974.

10. Seiden, A., Grossman, M.: Proposed Additional Guidelines for Hospital Care of Child Victims of Suspected or Alleged Rape. Appendix to P. Bart, et al., Hospital Subcommittee Report, included in *Recommendations and Report of Citizens' Advisory Council on Rape*, Cook Co., Ill., March 1975. (See Appendix V)

11. Seiden, A. M., Kellam, S.: by communication with the author.

12. De Francis, V.: *Protecting the Child Victim of Sex Crimes Committed by Adults*. The American Humane Association, Children's Division, P.O. Box 1266, Denver, Colorado 80201, 1966.

13. Katan, A.: Children Who were Raped, *Psychoanal. Study of the Child*, University Press, 28, 208-224, 1973.

CHAPTER

NINE

THE ROLE OF THE PSYCHIATRIST

It is only recently that rape trauma has been considered an area in which psychiatrists can make a significant contribution. The report of the Center for Women Policy Studies[1] addresses this issue directly:

> Acknowledging that there are lay persons who have much more experience with rape victims than most professionals and who have the personal characteristics to be better counselors, we nevertheless believe that the involvement of persons trained academically or clinically in crisis intervention, female sexuality, interpersonal relationships, and human behavior can be useful to a rape victim counseling program. They can serve as trainers, as consultants when problems arise, or as persons to whom a difficult case can be referred with confidence.

The potential role of the psychiatrist is multifaceted, with clinical, administrative, teaching, supervisory and research aspects.

There are presently urgent needs for education and training in a variety of settings. All hospital personnel assisting rape victims must become sensitized to the trauma of rape and the needs of the victim, and familiarized with basic concepts and methods of crisis intervention. The rapidly expanding body of knowledge about rape should be incorporated into the curricula of all medical schools and residency training programs, with continuing education efforts made at the local, state, and national levels of professional organizations. Criminal justice personnel are requesting training programs to improve their treatment of the victim, and community hospitals are requesting consultation to assist them in developing new services for victims. The Center for Women Policy Studies[2] recommends a coordinated and collaborative educational effort by physicians, hospitals, citizens, and criminal justice personnel as the most effective approach to training needs. The psychiatrist can have both direct and consultative roles in all of these educational endeavors.

The psychiatrist's role in a hospital crisis intervention program is briefly described in Chapter 4, and is discussed in some detail by McCombie, et al.[3] In the emergency room setting, the psychiatrist has a unique opportunity to function in a consultative capacity with a variety of other specialty services, as gynecology, pediatrics, nursing, and social work. A collaborative alliance must be established in the emergency room which will facilitate medical treatment, enable the victim to make a decision about legal involvement and help the victim work towards crisis resolution. Consultative/liaison skills are crucial to the development of integrated programs which depend on an interdisciplinary team approach as the primary treatment modality.

The psychiatrist can assist in the acquisition of skills necessary for taking an adequate history and assessing the victim's emotional state and ego functioning so that optimal treatment planning can occur. It is useful to differentiate short-term crisis goals from long-term issues which may require a psychiatric referral. The perspective of the psychiatrist in evaluating which victims might appropriately benefit from long-term intervention can be extremely important. Adequate case supervision with a focus on family dynamics and the life context in which the rape occurs should be a major concern of every program. Immediate or long-term supervision is especially important in hospital settings where counseling is often done by lay counselors or a variety of personnel functioning in training capacities.

The psychiatric service may be a referral resource for community agencies, crisis centers and other physicians. It is often the place to which people turn when there is a maladaptive or delayed response to the crisis. Patients with other emotional problems will also be rape victims, and this experience may intensify pre-existing problems. The psychiatrist, then, must be knowledgeable not only in the special issues around rape, but must have the clinical skills necessary to serve as consultant to other therapists, treatment team members, and community agencies.

There are a variety of situations in which the psychiatrist will have direct contact with the victim and/or family in the crisis period. The clinician may be working with a rape crisis program, or a victim experiences an acute grossly psychopathologic response to the rape, or a patient is raped during the time of ongoing psychiatric treatment. The following case report[4] illustrates the latter circumstance:

> A 27 year old married woman was raped during ongoing therapy. She had been treated for a schizo-affective schizophrenic disorder for two years with steady, although at times chaotic, improvement until the time of the rape. She had just separated from her husband of one year and had moved to a hotel. A few days later, a man posing as a maintenance person came to "inspect" the plumbing and raped her. She was appropriately angry and attempted to report the incident, an unusually assertive step for this woman. Since then, she has spent most of the year in mental hospitals and has been unable to deal either with the effects of the rape or with anything else. Out of the hospital, her self-abusive behavior rapidly triggers rehospitalization. She is markedly ambivalent about where to live, being too phobic to live alone or with strangers, and too incompatible to live with her husband. Most of the gains made in therapy have been lost, and her prognosis is most uncertain.

More often, the need for psychiatric services becomes apparent after the immediate crisis period, although this pattern may change as the climate of opinion about rape changes. Clinical experiences of most psychiatrists around rape-related issues have thus far been limited to cases in which a patient reveals a previous rape in the context of psychotherapy, as in the following vignette:[4]

> A single female presented for therapy at age 18 with symptoms of anxiety, depression and difficulties with friendships. Although bright and artistically talented, she was unable to concentrate at school, and worked instead at a succession of menial part-time jobs. During the first four months, the therapy rarely seemed to get off the ground or take any focus. If anything, she became more depressed and anxious. With the establishment of trust, the fact of a rape two years earlier emerged. The qualitative change in therapy sessions was dramatic, with steady growth in therapy since then. There has been significant improvement in the areas of self confidence, body image and interpersonal relationships.

This final case report[5] is presented in some detail to illustrate the complexities of the psychological, sexual and interpersonal sequelae of an earlier rape.

Mrs. B is a 36 year old married woman who was first seen with a chief complaint of "I have no interest in sex," a longstanding pattern for the 14 years of her marriage to Mr. B, who is 38 years old. The precipitant for seeking help was Mr. B's pressure at a time when their oldest daughter, age 12, was clearly approaching adolescence. They have one other child, a 10 year old son. Initially the history seemed unremarkable. They had both grown up in small midwestern towns in middle-class families which were described as very involved with the church and local community issues. They felt that neither of them had had much opportunity for discussion of sexuality in their families. Mr. B had his first sexual experience while he was in the army. He developed a gratifying relationship with a young woman who lived near his army base in Europe. He felt that he should marry the woman, but he was shipped out precipitously and although they corresponded, he never saw her again. He reported thinking more about her recently, with recurrence of the guilt he experienced at the termination of the relationship. Mr. B describes a few sexual partners subsequently, in the context of relationships. He met Mrs. B while he was a junior and she was a freshman in college, and they married following graduation. Mrs. B initially reported no sexual experience prior to marriage and blamed her inexperience for the sexual difficulties.

After meeting jointly for several sessions, Mrs. B requested an individual hour which Mr. B supported since he perceived that she was quite upset and was unable to communicate with him about the source of her anxiety. In the individual session Mrs. B reported a traumatic early experience. She had been on a trip to Europe with some high school classmates after graduation. On one occasion she decided to pursue some of her own interests instead of remaining with the group. She was accosted by two young men who initially seemed accommodating and helpful, offering to drive her to her destination. Instead they drove her to another place where they raped and left her, frightened, bruised and bewildered. She went to the police who disbelieved her story and sent her back to the group, making derogatory remarks about American women. She was extremely upset, but told no one because she felt guilty and ashamed. She felt that she should never have put herself in the position of being raped and that it was her fault. She did not seek medical help for the same reasons.

Upon returning home from the trip, she found that she was pregnant. She felt desperate and did not know where to turn. She finally sought out a high school classmate whom she had always seen as one of the "bad" girls, but who had once told her that she had had an abortion. The classmate helped her to arrange an illegal abortion in another city. The procedure was terrifying, dirty, and was accompanied by demands for intercourse by the abortionist. She went along with the sexual demands and had the abortion. She described feeling humiliated, degraded and vulnerable. She subsequently returned to school, depressed and confused, but again told no one. She lost 20 pounds in the two months following the abortion, developed insomnia and recurrent nightmares, and withdrew from social contacts. She did not date for the next year and when she started to date, she was unable to involve herself in a relationship that might involve any sexual activity.

When she met her husband she was interested in him because he did not push her sexually. They did not have intercourse until their marriage. She describes their sexual experiences as extremely traumatic for her. She felt angry, unresponsive and at times cried. She was not orgasmic. Her persistent feeling had been that she would rather have "nothing to do with sex" which was "disgusting." She would have flashbacks to her previous experiences and would "turn off." She described other aspects of her relationship with her husband as satisfactory, but she felt extremely uncomfortable about telling him of her past history.

She spent several months in individual therapy working on the events of the past, with a focus on her guilt and her difficulty dealing with her anger. She was subsequently able to tell her husband of her experience during a conjoint session. He felt shocked, deceived and conflicted, and alternated between supportive understanding and angry disbelief. The couple spent the next six months in therapy working on these issues with gradual improvement in their sexual relationship. They were able to share mutual reactions and understand the derivatives of the feelings they were experiencing, as well as the kinds of adaptive and defensive styles involved in coping.

The psychiatrist is in a particularly critical position in this kind of situation. He/she saw a couple presenting with what appeared to be a not uncommon kind of marital difficulty. When the past rape was revealed, it was essential that he/she was familiar enough with the problem to be able to put it in an appropriate perspective and help the patient work it through. It is not infrequent for the uninformed therapist to minimize the significance of this particular traumatic experience and to fail to see it as a determinant of presenting symptomatology. It is tempting to hark back to early developmental issues, which may indeed be factors in the response, or to question the patient's apparent "overreaction" either out of ignorance of the nature of the experience, or anxiety about one's own vulnerability or one's own aggressive or sexual fantasies or impulses. The therapist in this case was able to uncover and successfully work through a major crisis, using both individual and couples therapy models.

The role of the psychiatrist in research endeavors is described in the Epilogue.

References

1. Rape and its Victims: A Report for Citizens, Health Facilities and Criminal Justice Agencies, *The Response of Citizens' Action Groups: A Handbook*. Prepared by Center for Women Policy Studies, Washington, D.C., 1975. Copies available from Law Enforcement Assistance Administration, Washington, D.C.

2. Rape and its Victims: A Report for Citizens, Health Facilities and Criminal Justice Agencies, *The Response of Medical Facilities: A Handbook*. Prepared by Center for Women Policy Studies, Washington, D.C., 1975. Copies available from Law Enforcement Assistance Administration, Washington, D.C.

3. McCombie, S. L., Bassuk, E., Savitz, R., Pell, S.: Development of a Medical Center Rape Crisis Intervention Program. Presented at Special Session on Rape, American Psychiatric Association, Anaheim, California, 1975. Publication pending, American Journal of Psychiatry, April 1976.

4. Chappell, A.: by communication with the author.

5. Nadelson, C., Notman, M.: by communication with the author.

TOWARD THE FUTURE

The eradication of rape is contingent on educating and sensitizing our society to the meaning of the crime and the context in which it occurs. Innovative and empathetic services to victims will serve as a deterrent to the crime by facilitating reportage and thereby apprehension and prosecution of the assailants. Ultimately, however, the elimination of rape will require a massive reconsideration and restructuring of social values as well as a reorientation of the relations between the sexes. It is the thesis of this author that when the sex roles of both men and women are defined by individual needs and talents rather than by stereotypic expectations based on sex and power motives, only then will there be an end to rape.

In the meantime, there is much unknown and undone regarding the rape victim. What follows is a summary of the research questions which demand our attention:

1. What are the immediate and chronic effects of rape in the context of the victim's place in the life cycle, and what are the treatment implications?

2. What is the long-term impact of rape with regard to developmental aspects of sexual and aggressive drives?

3. What is the frequency of a prior rape among women who are currently psychiatric patients? Numerical and descriptive data may serve in part to clarify the first and second questions.

4. Can we identify and describe a sample of victims who do not come to public attention? All available information about victims is necessarily gleaned from a sample population which reports to either hospital or law enforcement. A survey of private physicians may provide information about a silent subgroup which may differ from the group who reports.

5. What is the impact of rape by a stranger as compared with rape by an assailant known to the victim?

6. Are there some women who are more vulnerable to rape victimization by virtue of personal factors? If so, identification of those factors which heighten vulnerability may be utilized for predictive and preventive measures.

7. Are some women subject to repeated rapes by virtue of living in neighborhoods and communities in which rape is endemic? If so, what can we learn of this subsample, and what measures are appropriate to prevent the crime and assist the victims?

8. What is the incidence of rape in children, as an issue separate from child abuse and incest? What are the immediate and long-term responses of children and their parents? Most authors cite that parental ability to work through the crisis determines the child's adjustment. It can be assumed, however, that children recognize the confrontation with sex and violence, and that this will have an impact separate from parental response.

9. What relevance does the sex of the clinician or counselor have on the outcome of treatment? Is it possible to predict whether certain victims will have a better response to one or the other sex?

10. What are the dilemmas posed for the male victim, with elaboration of both similarities and differences in the comparison with female victims?

11. As an allied issue, what is the prevalence of "battered women," that is, women who sustain chronic violent abuse by significant other males? While this issue is surfacing in media reports, it has received scant attention by professionals. In the past, such women have been labelled as "masochists" which may be an accurate but nonproductive assessment. Virtually nothing is known about this group, other than the suspicion that this may represent a considerably larger segment of the population than do victims of rape.

It is hoped that this monograph will generate a new level of interest and scrutiny, with psychiatrists assuming a leadership role in the investigation of issues concerning rape.

BIBLIOGRAPHY

This bibliography is presented only as a general guide to the wide variety of written materials currently available about the rape victim and should not be considered a comprehensive or definitive listing. Many of the sources listed contain their own useful bibliographies. The bibliography is divided into four sections which include books, reports, pamphlets and newsletters, and articles.

BOOKS

Amir, M.: *Patterns of Forcible Rape*. University of Chicago Press, Chicago, 1971

Brownmiller, S.: *Against Our Will: Men, Women and Rape*. Simon and Schuster, New York, N.Y., 1975

Burgess, A. W., Holmstrom, L. L.: *Rape: Victims of Crisis*. Robert J. Brady Co., Bowie, Maryland, 1974

Connell, N., Wilson, C., Eds.: *Rape: The First Sourcebook for Women*. New American Library, New York, N.Y., 1974

Horos, C. V.: *Rape*. Tobey Publishing Co., New Canaan, Connecticut, 1974

MacDonald, J. M.: *Rape: Offenders and Their Victims*. Charles C. Thomas, Springfield, Illinois, 1971

Medea, A., Thompson, K.: *Against Rape*. Farrar, Straus and Giroux, New York, N.Y., 1974

Russell, D. E. H.: *The Politics of Rape: The Victim's Perspective*. Stein and Day, New York, N.Y., 1975

Schultz, L. G., Ed.: *Rape Victimology*. Charles C. Thomas, Springfield, Illinois, 1975

REPORTS

Rape and Its Victims: A Report for Citizens, Health Facilities, and Criminal Justice Agencies. Prepared by Center for Women Policy Studies, Washington, D.C., April, 1975. The report consists of four separate handbooks: *The Police Response, The Response of Medical Facilities, The Response of Prosecutor's Offices,* and *The Response of Citizens' Action Groups*. Copies are available from Law Enforcement Assistance Administration, 666 Indiana Avenue, N.W., Washington, D.C. 20531

Report of the Public Safety Committee Task Force on Rape. Subcommittee of the District of Columbia City Council. City Hall, 14th and E Streets, N.W., Room 507, Washington, D.C. July, 1973

Report of the Task Force to Study the Treatment of Victims of Sexual Assault. County Council of Prince George's County, Maryland. March, 1973

The Crime of Rape in Denver. Prepared by Giacinti, T. A. and Tjaden, C. for the Denver High Impact Anti-Crime Program. Denver Anti-Crime Council, 1313 Tremont Place, Suite #5, Denver, Colorado 80204. 1973

Operation Rape Reduction Summary and Recommendations of the National Rape Reduction Workshop. Prepared by Sheppard, D. I. and Farmer, D.J., Denver Anti-Crime Council, 1313 Tremont Place, Suite #5, Denver, Colorado 80204. 1973

Report of the Subcommittee on the Problem of Rape in the District of Columbia. Mental Health Committee, The Medical Society of the District of Columbia Medical Annals of D.C., 41, 703-704, November 1972

Rape Victimization Study. Prepared by Queen's Bench Foundation, 1231 Market Street, Suite 15, San Francisco, California 94102, June 1975

PAMPHLETS AND NEWSLETTERS

Counselor's Manual. Rape Crisis Center, P.O. Box 1312, Madison, Wisconsin 53701. November 1973

Feminist Alliance Against Rape Newsletter. P.O. Box 21033, Washington, D.C. 20009

How to Start a Rape Crisis Center. Rape Crisis Center, P.O. Box 21005, Kalorama Street Station, Washington, D.C. 20009. August 1972

Medical and Legal Aspects of Rape. Women Organized Against Rape, P.O. Box 17374, Philadelphia, Pennsylvania 19105. 1973

Rape Prevention Tactics. Rape Crisis Center, P.O. Box 21005, Kalorama Street Station, Washington, D.C. 20009

Stop Rape. Women Against Rape, 2445 W. 8 Mile, Detroit, Michigan 48203

What the Rape Victim Should Know About the "Morning After" Pill, prepared by Weiss, K., Advocates for Medical Information, 2120 Bissonnett, Houston, Texas 77005, 1975

ARTICLES

Bard, M., Ellison, K.: Crisis Intervention and Investigation of Forcible Rape. *The Police Chief* 41:5, 68-73, May 1974

Bart, P.: Rape Doesn't End With a Kiss, *Viva* 2:9, 39-42, 100-101, June 1975

Breen, J. L., Greenwald, E., Gregori, C.: The Molested Young Female, Evaluation and Therapy of Alleged Rape. Symposium on Pediatric and Adolescent Gynecology, *Pediatric Clinics of N.A.* 19:8, 717-722, August 1972

Burgess, A. W., Holmstrom, L. L.: The Rape Victim in the Emergency Ward. *American Journal of Nursing* 73:10, 1740-1745, October 1973

Burgess, A. W., Holmstrom, L. L.: Rape Trauma Syndrome. *American Journal of Psychiatry* 131:9, 981-86, September 1974

Burgess, A. W., Holmstrom, L. L.: Coping Behavior of the Rape Victim. *American Journal of Psychiatry* 133:4, 413-418, April 1976.

Capraro, V. J.: Sexual Assault of Female Children. *Annals New York Academy of Science* 142, 817-819, 1967

Cohn, B. N.: Succumbing to Rape? *The Second Wave*, 2:2, 24-30, 1972

Densmore, D.: On Rape, No More Fun and Games. *A Journal of Female Liberation* 6, 57-84, May 1973

Evrard, J.: Rape: The Medical, Social and Legal Implications. *American Journal of Obstetrics and Gynecology* 111:2, 197-199, September 1971

Griffin, S.: Rape: The All-American Crime. *Ramparts* 26-35, September 1971

Greer, G.: Seduction is a Four Letter Word. *Playboy*, January 1973

Hayman, C. R., Lewis, F. R., Stewart, W. F., Grant, M.: A Public Health Program for Sexually Assaulted Females. *Public Health Reports* 82:6, 497-504, June 1967

Hayman, C. R., Lanza, C., Fuentes, R., Algor, K.: Rape in the District of Columbia. *American Journal of Obstetrics and Gynecology* 113:1, 91-97, May 1972

Hayman, C. R., Lanza, C.: Sexual Assault on Women and Girls. *American Journal of Obstetrics and Gynecology* 109:3, 480-486, February 1971

Herschberger, R.: Is Rape a Myth? *Adams Rib*, Chapter 3. Harper and Row, New York City, 1948, (HAR/ROW ed., 1970)

Hibey, R. A.: The Trial of a Rape Case: An Advocate's Analysis of Corroboration, Consent, and Character. Reprinted in *Rape Victimology*, ed. Leroy G. Schultz, Charles C. Thomas, Springfield, Ill., 1975 (originally published in *American Criminal Law Review* 11:2, Winter 1973).

Kanin, E. J.: Selected Dyadic Aspects of Male Sex Aggression. Reprinted in *Rape Victimology*, ed. LeRoy G. Schultz, Charles C. Thomas, Springfield, Illinois, 1975. (*Journal of Sex Research* 5:1, February 1969).

Katan, A.: Children Who Were Raped. *Psychoanalytic Study of the Child*, University Press 28, 208, 1973

Komisar, L.: Violence and the Masculine Mystique. Reprinted by KNOW Press, P.O. Box 86031, Pittsburgh, Pa. 15221, 1973

Margolin, D., Sheldon, A.: Rape: The Experience, Rape: The Facts, Rape, A Solution, *Women, A Journal of Female Liberation* 3:1, 18-23, 1972

Massey, J. B., Garcia, C. R., Emich, J. P.: Management of Sexually Assaulted Females. *Obstetrics and Gynecology* 38:1, 29-36, July 1971

McCombie, S. L., Bassuck, E., Savitz, R., Pell, S.: Development of a Medical Center Rape Crisis Intervention Program. *American Journal of Psychiatry* 133:4, 418-421, April 1976

Mehrhof, B., Kearon, P.: Rape: An Act of Terror. *Notes From The Third Year: Women's Liberation*, 79-81, 1972

Menen, A.: The Rapes of Bangladesh. *Reflections*, 8:5, 6-14, 1973

Metzger, D.: It is Always the Woman Who is Raped. *American Journal of Psychiatry* 133:4, 405-408, April 1976

Notman, M. T., Nadelson, C. C.: The Rape Victim: Psychodynamic Considerations. *American Journal of Psychiatry* 133:4, 408-413, April 1976

Rich, A.: Caryatid, A Column. *American Poetry Review*, May/June 1973

Rush, F.: The Sexual Abuse of Children: A Feminist Point of View. Section 3 of *Rape: The First Sourcebook for Women*, Connell, N. and Wilson, C., Eds., New American Library, New York, N.Y., 1974

Schmidt, P.: Rape Crisis Centers. *Ms. Magazine*, 14-18, September 1973

Schultz, L. G., De Savage, J.: Rape and Rape Attitudes on a College Campus. Chapter 6, *Rape Victimology*, ed. LeRoy G. Schultz, Charles C. Thomas, Springfield, Ill., 1975

Schurr, C.: Rape: Victim as Criminal, Pittsburgh Forum, 1971, reprinted by KNOW Press, P.O. Box 86031, Pittsburgh, Pa. 15221

Schwendinger, J. R., Schwendinger, H.: Rape Myths: In Legal, Theoretical and Everyday Practice, *Crime and Social Justice*, Spring-Summer 1974

Suspected Rape. ACOG Technical Bulletin Number 14, July 1970 (Revised, April 1972)

Sutherland, S., Scherl, D.: Patterns of Response Among Victims of Rape. *American Journal of Orthopsychiatry* 40:3, 503-511, April 1970

Symonds, M.: Victims of Violence: Psychological Effects and After-Effects. *American Journal of Psychoanalysis* 32:1, 1975

Wasserman, M.: Rape: Breaking the Silence. *The Progressive*, 19-23, November 1973

Weis, K., Borges, S. S.: Victimology and Rape: The Case of the Legitimate Victim, *Issues in Criminology* 8:2, 71-115, Fall 1973

Weiss, R. J., Payson, H. E.: Gross Stress Reaction I in A.M. Freedman and H. I. Kaplan (eds.). *Comprehensive Textbook of Psychiatry*, Baltimore, Williams and Wilkins, 1027-1031, 1967

Wood, P. L.: The Victim in a Forcible Rape Case: A Feminist View. Reprinted in *Rape Victimology*, ed. LeRoy G. Schultz, Charles C. Thomas. Springfield, Ill., 1975 (originally published in *American Criminal Law Review*, 11:2, Winter 1973)

APPENDIX I

(*Reprinted by permission of Dr. Talbert and Mr. Warren*)

GUIDELINES FOR MANAGEMENT OF SUSPECTED RAPE[1]

Luther M. Talbert, M.D.
Department of Obstetrics and Gynecology
University of North Carolina School of Medicine
Chapel Hill, North Carolina

and

Professor David G. Warren
Institute of Government
University of North Carolina
Chapel Hill, North Carolina[2]

[1] Derived in part from ACOG Technical Bulletin No. 14, as revised April, 1972.
[2] With the assistance of Ms. Jean Winborne Boyles, Police Attorney, Town of Chapel Hill, N.C.; Dr. Janet J. Fischer, Professor, Department of Medicine, Division of Infectious Disease, University of North Carolina School of Medicine, Chapel Hill, N.C.; and Arthur J. McBay, Ph.D., Chief Toxicologist, Office of the Chief Medical Examiner, Chapel Hill, N.C.

Department of Obstetrics & Gynecology
School of Medicine
University of North Carolina
Chapel Hill, N.C.
27514

AUTHORIZATION FOR
EMERGENCY RAPE EXAMINATION

I hereby give permission to Dr. _____ to perform
a complete physical examination including pelvic examination and for
that physician to collect appropriate specimens of vaginal secretions,
blood, hairs and other related specimens for laboratory examinations.
I also consent to treatment related to my present condition, including
penicillin and/or other antibiotics.

_____   _____Age:_____
Date   Patient's Signature (in all cases)

_____   _____
Date   Guardian's Signature (if appropriate)

_____   _____
Date   Witness

Department of Obstetrics & Gynecology
School of Medicine
University of North Carolina
Chapel Hill, N.C.
27514

AUTHORIZATION FOR
RELEASE OF RAPE INFORMATION AND SPECIMENS

Release: I authorize any pertinent information obtained by the
physician from the history and physical examination of the
patient to be released to appropriate law enforcement
officials and in addition that the following specific items
may be so released:

_____

_____

_____

_____

| _____ | _____Age:_____ |
| Date | Patient's Signature (in all cases) |

| _____ | _____ |
| Date | Guardian's Signature (if appropriate) |

| _____ | _____ |
| Date | Witness |

Receipt: I certify that I have received the items listed above from

_____ on _____ at _____.
                              (date)   (time)
All items were properly identified.

| _____ | _____ | _____ |
| Signature | Title | Date |

| _____ | _____ |
| Witness | Date |

GUIDELINES FOR MANAGEMENT
OF SUSPECTED RAPE[1]

Luther M. Talbert, M.D. and David G. Warren

The following document is intended to serve as a general guideline for the management of possible victims of sexual assault who present themselves at the Emergency Room; it is not intended to be a binding set of rules. The situational needs for modification of these guidelines cannot be determined in advance, and the physician caring for the particular patient must exercise practical and ethical judgments within the framework of good medical care, due regard for privacy and a special sense of social responsibility.

While the legal rights of the patient, the physician and other parties must be protected, it cannot be stressed too strongly that the primary role of the physician and other hospital personnel is the care of the patient, including protection of her physical, emotional and social welfare.

A. LEGAL DEFINITIONS

1. Rape is carnal knowledge of a female aged 12 or more by force and against her will. Actual physical force is not necessary since submission under fear or duress takes the place of physical force. Consent by the woman, voluntarily given, is a complete defense.

2. Carnal knowledge of a female under the age of 12 is rape regardless of her consent and regardless of no use of force.

3. "Statutory rape," a lesser crime, is carnal knowledge of a girl between the ages of 12 and 16 who was previously a virgin, regardless of her consent and regardless of no use of force.

4. Carnal knowledge is the slightest penetration or entry into the vulva by the penis.

---

[1] Derived in part from ACOG Technical Bulletin No. 14, July, 1970.

   5. Child abuse is physical injury (including sexual acts) to a child under sixteen caused or allowed by any parent or caretaker of that child.

B. CONSENT
The examination of rape victims will ordinarily be considered an emergency and the patients' consent will usually be implied from the situation. However, to protect both the physician and the patient, witnessed consent for the following procedures should be obtained in writing, if possible. If oral, a note should be entered in record describing who gave consent, who was present, and why written consent was not obtained.

   a. Physical examination
   b. Collection of specimens and other materials
   c. Photographs (if such are to be obtained)
   d. Release of information and specimens to law enforcement officials
   e. Treatment of present condition, including administration of penicillin for possible venereal disease.

Appropriate consent forms should be completely filled in at the time consent is obtained. The patient's signature should always be obtained, regardless of her age. If she is physically, mentally or emotionally incapacitated, examination and treatment should proceed as for other emergency conditions, without written consent. If the patient is under 18, reasonably diligent attempts should be made to notify her parents or guardians, unless she clearly objects; necessary medical examination and treatment, however, should not be delayed pending notification.

C. PHYSICIAN'S RESPONSIBILITY
The physician must protect the medical and personal interests of the patient, but must recognize his own social responsibilities relating to criminal justice. Every "rape" incident is a potential criminal court case. The examining physician might be asked or subpoenaed to court to testify about what actions he took, what findings he made and what records he kept. Whether rape occurred is a legal matter for court decision and is NOT a medical diagnosis. The physician's responsibility is to assist the court by describing

accurately the observations he made and not to draw a conclusion about whether the evidence constitutes a rape. He may, however, be asked his opinion about whether there had been carnal knowledge of the patient.

There is no legal requirement for the hospital or physician to report apparent crimes to the police. Once the police begin an investigation of a crime, the hospital and physicians are then required to cooperate and to furnish information which is not protected by rules of confidentiality. The medical portions of a patient's record are confidential and can be released only with the patient's consent.

D. Patient's Responsibility

The patient can at any time report the incident to the police without filing an official complaint and without commencing a police investigation. Also, the patient can, at any time, file an official complaint and commence a police investigation. The physician can assist both the patient and the police by facilitating her contact with police:

(a) The physician or counselor will ask the patient if she has any objection to reporting the incident to police, regardless of whether she desires now or later to file an official complaint. The police can be requested to keep the report confidential and not to begin an investigation at this time. If no objection, give the following information to the Hospital Administrator for reporting to the police:

Name _____ Age _____
Where occurred _____
Time occurred _____
Report to be kept confidential? _____

(b) Ask patient if she desires now to talk with the police or to file an official complaint with the police in order to start a police investigation of the incident. If so, make arrangements for the patient to be interviewed by the police at a time and place agreeable to the patient. If not now, suggest that the hospital counselor will talk to her further later about the matter of talking to the police or filing a complaint.

E. Summary of Protocol

The examining physician should follow this guide:

a. Get consent (if conditions permit)
b. Get history in patient's words and do physical examination
c. Treat patient's condition
d. Record history and examination findings
e. Get laboratory work and collect specimens for possible later use by police
f. Ask patient if police should be contacted
g. Notify police if patient desires to report incident
h. The persons to contact with law enforcement authorities are as follows:

Chapel Hill Police	929-2121
Carrboro Police	942-8537
Orange Co. Sheriff	942-6300
Durham Police	688-8251
Durham Co. Sheriff	682-2167

i. Protect against disease, pregnancy and psychic trauma

F. Detailed Discussion of Medical Procedures

1. *History*

A good history must be obtained and written down as quotations in the patient's words. The time, place and circumstances should be recorded. The patient's emotional state should also be noted (e.g., hysterical, inebriated, stoic). Has the patient taken a bath since the alleged assault?

2. *Examination*

After obtaining written consent to examine the patient, the physician should proceed promptly with the physical examination. There is no requirement whatsoever for a court order or police notice in order to carry out a medical examination on the possible victim of a rape. Great care must be exercised not to destroy any evidence which may later be useful in a police investigation or which may contribute to establish guilt or innocence in a case of suspected rape. While a general physical examination must be done and be recorded, the following should be stressed.

a. General appearance: bruises, lacerations, torn or bloody clothing and condition of patient (e.g., inebriated, hysterical, punch-drunk) should be recorded.

b. External genitalia: evidence of trauma.

c. Speculum examination: inspect cervix and vagina with a non-lubricated speculum. Vaginoscope (available in Ob-Gyn Clinic) may be necessary in children.

d. When the vagina has not been penetrated in prepubescent children, it may not be possible to perform a vaginal examination. If the vagina has been penetrated in a prepubescent child, penetration will usually be evidenced by perineal lacerations. In this case, it is extremely important that the upper vagina be inspected for lacerations, even though active bleeding is not present at the time. If this cannot be done easily in the emergency room, the patient should be anesthetized.

3. *Laboratory Specimens*

a. Collection

Laboratory investigation should include the following: (1) A drop of vaginal secretions and a drop of aspirate from the endocervix are placed on a slide and covered with a cover slip. This is inspected fresh under microscope for living sperm and if living sperm are present, so record. (2) Obtain a specimen from the posterior fornix with a cotton-tip applicator. Smear this on two clean slides exactly as if for a Papanicolou smear and let the specimen air-dry. It will last indefinitely. (3) Put the above swab in a dry, screw-top test-tube. In addition, using a glass aspirating bulb, collect all possible vaginal secretions and put them in the same dry tube with the cotton swab. *Do not* add saline, formalin, or any other liquid to the tube. (4) Spread a paper towel under the patient and comb the pubic area lightly for loose hair. Place comb on towel and fold towel with comb and combings inside. Place towel in a plastic bag and seal completely with tape (for foreign hair, etc.). (5) *Pull out* five or six pubic hairs and place in a plastic bag and seal (known sample).

b. Disposition of Specimens

All specimens should be tagged with patient's name and unit number, the date, and the name of the person having custody at each stage of handling the specimen. The tag should be signed by each person prior to handing it to another. The air-dried slide, the tube containing the posterior fornix specimens, and other specimens should be handed to the police if the police are involved but only after the patient has signed the appropriate authorization form and the custody tag has been signed. A receipt on the form should be obtained from the policeman indicating that he has received the specimen. It then becomes the responsibility of the law enforcement official to transfer the speciman to the SBI Offices in Raleigh, where thorough laboratory studies, including semen-typing, acid phosphotase, etc., will be done by the SBI. If police are not immediately involved, the same tagging procedure should be followed and the specimens placed in the freezer. If the police are not involved within six months, the specimens shall be destroyed. On some occasions the police may ask the hospital to supply blood specimens or other special specimens from the victim. Obtain the consent of the patient on the consent form and proceed.

4. *Clothing and Other Material*

These may be useful as physical evidence. Therefore, care should be taken to preserve such material, either when the police request it or when it is obviously evidentiary, if the patient agrees. If the patient has filed an official complaint and the police have begun an investigation, an authorization form should be obtained for clothing, photographs and any other physical evidence the patient agrees to submit. They should be turned over to police authorities in return for a detailed receipt, as provided in the form.

5. *Objective Statements*

The record should contain the patient's statement. It should give descriptions of the physician's findings and what he did. It should state to WHOM he delivered specimens, clothing or photographs. NEVER write in the record your opinion concerning whether or not the patient was raped. The phrases "suspected rape" or "alleged rape" may be used when necessary. The physician should remember that both he and the record may be subpoenaed and that he may be required to testify. All information should be exact and sufficiently detailed to avoid any misinterpretation. Negative findings are as important as positive ones and may assist in the protection of a person who has been falsely accused. The victim or her parent or legal guardian should be given a full account of the medical results of the examination. The physician may, for example, say that he did find sperm in her vagina but he is not at liberty to go from this to say the patient was legally raped. In the case of a pre-pubescent child he may inform the parents or guardians that there are no lacerations and no evidence of trauma to the vaginal or perineal structure; but if such exists it is his obligation, as with any other patient, to inform the parents. On the other hand, he is not at liberty to conclude that a laceration resulted from rape (a bicycle handle-bar may produce an identical injury).

6. *Care of the Child*
Children under the age of 16 are the direct responsibility of the Pediatrics Department. (See accompanying document.) Note that the N.C. Child Abuse Law requires any professional person who has reasonable cause to suspect that any child has been abused or neglected by a parent to report the case to the Orange County Director of Social Services. Therefore, if such a case is not transferred to Pediatrics then the report should be made directly.

7. *Prevention of Disease*
The "attacker" may have a venereal disease. Ask the patient about the possibility of any penicillin allergy. If you determine that there should be no penicillin reaction, the patient should be given 4.8 million units of aqueous procaine penicillin IM 30 minutes after 1 gm. of oral probenecid.

The risk of contracting syphilis is very small (around 0.1%) and prophylaxis is probably not indicated. However, an STS should be drawn initially and, along with examination, should rule out established syphilitic infection. If penicillin is used for gonorrhea phophylaxis, it will abort any syphilis acquired at the same time. However, it is insufficient to treat established syphilitic infection which should be adequately treated if diagnosed. *If* spectinomycin is used as prophylaxis for gonorrhea, an STS must be done monthly for three months to be sure that syphilis does not develop. Spectinomycin may mask incubating syphilis. Also spectinomycin should not be given to children or to pregnant women. Its use should probably be restricted to patients who cannot receive penicillin. It is advisable to see the patient six weeks later to obtain cervical cultures for G.C. and a Serologic test for syphilis. Appropriate appointments should be made in the gynecology or pediatric clinic.

8. *Prevention of Pregnancy*

The possibility of pregnancy should always be considered. There is no single procedure unifomly accepted at this time. Stilbestrol 25 mg. daily by mouth for 5 days or Premarin 25 mg. daily for 5 days has been recommended. If a menstrual period does not begin within one week after the expected onset, a D & C should be recommended to the patient. If pregnancy occurs as a result of rape, abortion should be recommended to the patient.

9. *Management of Psychic Trauma*

All patients over the age of 16 will be seen by a counselor as part of the Emergency Room treatment of rape. It will be the responsibility of the counselor to deal with the psychological and social problems resulting from the incident. The counselor will also have legal information should the patient wish to discuss this.

10. *Follow-up of Patient*

All patients should be followed to be certain that they do not develop a venereal disease or become pregnant. Every patient should be given at least one follow-up appointment in the appropriate gynecology or pediatric clinic.

APPENDIX II

GUIDELINES FOR CARE OF THE VICTIMS OF RAPE,
SEXUAL ASSAULT
N.C. MEMORIAL HOSPITAL
EMERGENCY ROOM

(Reprinted with permission of the Emergency Room Rape Crisis Program, North Carolina Memorial Hospital, Chapel Hill, N.C.)

GUIDELINES FOR CARE OF THE VICTIMS OF RAPE, SEXUAL ASSAULT

N.C. MEMORIAL HOSPITAL
EMERGENCY ROOM

Rape, one of the major violent crimes in the U.S., affects the lives of thousands of women each year. The reaction of the victim may be expressed in many different ways—anger, fear, anxiety.

The hospital system places additional demands on the victim—to answer questions and describe the details of the rape. For many women, the hospital experience may be almost as traumatic as the assault. On arriving at the emergency room, the victim is caught up in the day-to-day workings of the hospital, in a process which is routine to the ER staff but new and not understood by her.

Therefore, it is important to give the victim an environment of safety, empathy, and confidentiality. Equally important is that the victim feels she is in control of her ER process—that she will decide what happens to her. The ER staff is in a position to have a major impact on the experience of the rape victim.

ADMISSION OF THE PATIENT
The ER secretary will quickly and discreetly obtain the victim's name, address, and date of birth. The secretary will immediately notify the triage nurse that there is a "7273" (RAPE) patient. The victim will *not* be asked to sit in the public waiting area.

MEDICAL TEAM

Triage Nurse
1. respond to secretary's coded call immediately.
2. escort victim to family room or other private area.

3. consult with victim regarding whom she would like to stay with. (It may be a person who came with her to the hospital or a nurse or she may prefer to be alone.)

4. assign a nurse to either stay with the victim or check on her at frequent intervals.

Nurse Assigned to Patient

1. make victim feel that she is in a safe environment.

2. encourage her to express her feelings.

3. prepare her for people (counselor, physician, etc.) and events (examination, etc.) that she will encounter in the ER.

4. communicate with triage nurse as soon as possible to make decision regarding order of victim's care (i.e., physician, counselor, law enforcement officer).

5. notification of appropriate persons.

example:
A woman comes into the ER in an extremely upset state and doesn't seem in any condition to tolerate a pelvic exam. In this case, a counselor would spend time with the victim until she is more able to tolerate the medical exam.

example:
A victim presents to the ER with profuse vaginal bleeding and is quite anxious. The counselor may stay with the victim during the examination and wait until after the physician has completed the exam and treatment to talk with the patient.

example:
A woman calls the ER needing to talk with someone regarding a rape. The secretary should obtain the victim's phone number and have the counselor return the call.

6. prepare examining room.

7. place specimen kit in examining room.

8. attach record packet to chart.

9. communicate appropriate information to physician and counselor.

10. let victim know 1) what is expected of her during physical exam, 2) what doctor will do, and 3) why the procedures are necessary.

11. give the victim as much emotional support and make her as physically comfortable as possible.

12. assist the physician as well as the victim during the examination.

13. have a working knowledge of the specimens to be obtained (see section of laboratory specimens in Guidelines for *Management of Suspected Rape*).

14. obtain fingernail scrapings.

15. give specimens and clothing to be used as evidence to hospital security guard.

16. administer medications for prevention of pregnancy and V.D. if ordered and explain possible side effects.

17. provide measures for cleanliness after the exam (shower on CRU, toothbrush, etc.).

18. give follow-up wound care instructions.

19. make return appointments and referrals.

20. help in obtaining transportation from hospital.

21. log victim's number in appropriate places.

Physician

After assessing the victim's physical and emotional status, the nurse will make judgments regarding the order of care. Depending upon the victim's needs, the Gyn physician will be notified that there is a rape victim in the ER and called again when the victim is ready to be examined. The resident to be involved in this care is the second or third (chief) resident on call—not the first year resident. Private patients will be seen by the attending physician on call or the patient's personal physician.

The physician, counselor, and nurse will communicate and collaborate in the best interest of the victim.

The following is a brief guide for management of rape victims. For further details for management of rape victims, see the document entitled MANAGEMENT OF SUSPECTED RAPE which is on file in the Emergency Room.

I. If circumstances permit, have the victim sign the following documents:

(a) authorization for emergency rape examination and collection of pertinent specimens, and

(b) authorization for release of rape information and specimens.

The victim may not wish to sign document (b). This document does not have to be signed for the physician to examine and counsel her about and/or treat her for possible exposure to venereal disease and pregnancy.

II. Record a complete history of the incident and a complete physical examination. Record especially any evidence of trauma. The presence of a nurse is required for the interview (i.e. history taking) and the physical examination.

III. At the time of the pelvic examination, these procedures for the collection of evidence should be followed:

(a) Comb the pubic hair onto a clean paper towel using a *new comb*. Roll the comb and the hair in the paper towel and place in a plastic bag supplied by the nurses. Label this "pubic hair combing."

(b) Pluck out five or six pubic hairs and place in a plastic bag and seal. Label this "victim's pubic hair."

(c) Insert a speculum and aspirate a drop of secretions from the endocervix, place them on a slide with saline and cover with a cover slip. Inspect this fresh for sperm and if sperm are present, so record in chart.

(d) Obtain a specimen from the posterior fornix with a cotton tip applicator. Smear this on two clean slides exactly as if for a Papanicolaou smear and let the specimen air dry. Place in a container supplied by the nurse and label with victim's name and unit number.

(e) Drop the swab used in (d) above into a dry screw-top test tube. In addition, using a glass aspirating bulb, collect all possible vaginal secretions and put them in the same dry tube with the cotton swab.

(f) Obtain GC culture—2 sterile cotton swabs on 2 Thayer Martin plates.

(g) Initial (doctor's and nurses's initials) each piece of evidence.

This completes the collection of specimens for evidence. All specimens should be placed in a bag supplied by the nurse, tagged with the victim's name and unit number, and handed to either the nurse or one of the security officers of the hospital. The chain of evidence document should be appropriately signed. If the police are present and if the victim wishes to report the incident to the police, the specimens may be given directly to the police officer after he signs the appropriate receipt.

IV. *Prevention of Disease.* A serology should be drawn to rule out established syphilitic infection. If the victim is not allergic to penicillin, she should be given 4.8 million units of aqueous procaine penicillin intramuscularly 30 minutes after one gram of oral probenecid. If she is allergic to penicillin, spectinomycin may be used.

V. *Prevention of Pregnancy.* The victim should be given one of the postcoital preparations for pregnancy if indicated. Stilbestrol 25 mg. daily for 5 days or Premarin 25 mg. daily for five days may be used.

VI. *Management of Psychic Trauma and Follow-Up.* Victim should be given the opportunity to make contact with one of the rape counselors. A follow-up Gyn examination should be scheduled at an appropriate time.

Counselor

The timing of intervention and care by the counselor, too, will be dependent upon the individual victim's needs.

1. The counselor is responsible for primary counseling before, during, and/or after the physical exam as indicated.

2. The counselor will spend some time with the victim after the exam for continued counseling and will give her a booklet on rape. This booklet will also be shared with the family members or friends who accompany her as indicated. This time may be used to discuss appropriate source of referral.

3. The counselor will telephone the victim within 48 hours following her visit to the ER to discuss feelings, questions, problems, etc.

4. With the consent of the victim, the counselor will make a "blind report" to the appropriate law enforcement agency if the victim does not wish to become directly involved with legal proceedings. The report and consent forms are attached to the victim's chart. This report should be called in after the victim has left the ER in order to decrease the victim's anxiety and avoid confrontation with the police in the ER.

As part of the medical team, all information communicated to the counselor will be considered privileged.

ON-CALL SCHEDULE
Daytime: If a victim comes in during the day (8:00 a.m.-4:30 p.m.), one of the following can be called:

(Names and phone numbers are listed.)

Evenings and Weekends: (4:30 p.m.-8:00 a.m.): The counselors' call schedule will be kept in the ER in the on-call schedule book along with pertinent phone numbers. A sign-in book will be maintained in the ER telling where the counselor on-call can be reached. These numbers will not be given out. If a victim should call wanting to speak with a counselor, the secretary will have the counselor return the victim's call.

POLICE AND HOSPITAL SECURITY
Should law enforcement officers accompany the victim to the ER, or already be in the area, the nurse or secretary will inform officers that the victim is in the ER primarily for medical and emotional care. If the victim requests their presence in the ER, the officer will remain in the *waiting area.* If the victim wishes to talk with law enforcement officers when medical care is completed, the ER secretary will notify the officer that he is needed to return. The law enforcement officer will come to the ER to talk with her or take her to the police station if she wishes.

Specimens obtained during the physical exam and clothing to be used as evidence will be given to the hospital security guard for storage.

Law enforcement officers should *not* be allowed in the examining room.

TRANSPORTATION
Part of the total care of the victim is to provide her with a safe means of transportation to another environment of security.
Resources:
1. friend or relative of victim.
2. social worker on-call (may release money if needed)
3. Chapel Hill-Carrboro Rape Crisis Center—967-RAPE

FOLLOW-UP

The ER secretary will give an appointment to the victim for return to the GYN Clinic as specified by the doctor. At the time of the clinic visit, a questionnaire will be given to the victim to help evaluate the care she has received. If she should not return for her clinic visit, specified clinic nurses will see that a questionnaire is mailed to her.

RECORDS

A packet of information and forms pertinent to the care of the rape victim will be kept in the ER to be attached to the chart. After the ER procedures have been completed, the packet will be detached and kept in a locked file. The chart will be returned to Medical Records in the normal manner. The psychiatrist in charge of supervising the counselors will review these charts when appropriate. The chart of each rape victim will be assigned a special number which will be put on the victim's and counselor's questionnaire and kept on file in the ER for further reference by the counseling program.

APPENDIX III

FREQUENT REACTIONS TO RAPE

Chapter Two of *Information and Guidance for Adult Victims of Rape*, Boyles, J., Cole, K., Donadio, B., Hilberman, E., Peace, J., Reice, T. Prepared for Emergency Room Rape Crisis Program, North Carolina Memorial Hospital, 1974.

(Reprinted with permission of the Emergency Room Rape Crisis Program, North Carolina Memorial Hospital, Chapel Hill, North Carolina.)

Every woman reacts to rape in a different way. In talking with rape victims it has been found that most women would appreciate knowing what kinds of reactions and feelings other women have had to rape, so that they can be better prepared to handle their own feelings.

This note is not an attempt to tell you exactly how you will react to rape. These are just some suggestions for you to think about so that you can recognize that the different feelings you might have are normal, and that you don't have to think about them alone. A counselor from our program can help you find someone with whom you would like to talk.

In a rape the woman has had her safety and her life seriously threatened and disrupted. She has been subjected to a humiliating personal violation as well as to physical danger. She has been the victim of a violent crime, and she has temporarily lost her sense of control over her own fate. Nonetheless, she will somehow face this incident and its effects. This crisis, like any other, can be overcome and the woman can go on to lead a happy and satisfying life. The rape need not haunt her for a lifetime.

The immediate reaction is usually anxiety, disbelief and fear. Every person reacts to stress in a different way. Some women respond with crying, sobbing, shaking, and restlessness, while others appear outwardly calm and controlled. Going to a hospital emergency room for health care or talking with police about the crime can be upsetting to the woman. It's important to know you don't have to go through this alone. A family member, a friend, a counselor from our program can be with you.

In the days and weeks which follow, there are a number of things the victim may be worried about: Should she tell her family, husband, boyfriend, fiancee, or friends what has happened? Will friends and neighbors find out? Will she become pregnant? Should she press legal charges? Will she be able to identify the rapist? Should she talk to her clergy? Will there be publicity in the newspapers? At a time when there are different choices and decisions to be made, the victim may not be able to think clearly. It is normal to be worried, and to have trouble concentrating. Again remember you don't have to go through this alone; you can call or meet with a counselor to talk about these concerns.

The immediate reaction to rape is over when the woman becomes less anxious and returns to her daily life routine: to her job, family responsibilities, school, etc. This may occur in days or weeks depending on the victim and the circumstances. During this time the victim will outwardly appear to herself, close family, and friends to be making a good adjustment. In order to go on with her usual life, she may often deny her strongest reactions to the rape. While refusing to think about her feelings, the victim may still be aware of the incident through nightmares or daydreams. These reactions seem to be necessary to the victim to give her the feeling she has lived through the incident, and in time, will adjust.

As more time passes, the woman's awareness of what has happened to her as the result of the rape increases, and the feeling that "everything is back to normal" often gives way to depression. She may spend a lot of time reliving the incident and having feelings of anger, guilt, fear, revenge, self-blame. This kind of depression is a normal reaction to rape, and part of the process of understanding and resolving what has happened to her. The woman feels a strong need to talk about her feelings about herself, her assailant and her world. She may also feel the need to get away, wanting to change her address or limit her activities. She may feel fear for her personal safety, and may be very uncomfortable in situations which remind her of the incident: being indoors, outdoors, alone, in crowds, having someone walk behind her. The guilt and self-blame some victims feel may be the way a woman sees the incident because of her own need to be in control of her life. If the woman can think she had control over the rape, then she can have the false feeling that this won't happen to her again in the future. Many women have found it helpful at this time to get help from a counselor to talk about these feelings. After overcoming the fear and anger, the woman will probably feel greatly relieved. She comes to see herself as a worthwhile person who has been unjustly and unfairly wronged.

APPENDIX IV

A NOTE TO THOSE CLOSEST TO RAPE VICTIMS:
FAMILIES, SPOUSES AND FRIENDS

Chapter Seven of *Information and Guidance for Adult Victims of Rape*, Boyles, J., Cole, K., Donadio, B., Hilberman, E., Peace, J., Reice, T. Prepared for Emergency Room Rape Crisis Program, North Carolina Memorial Hospital, 1974 (originally published by the D.C. Rape Crisis Center who graciously consented to our adaptation for The North Carolina Memorial Hospital Use).

(Reprinted with permission of the Emergency Room Rape Crisis Program, North Carolina Memorial Hospital, Chapel Hill, North Carolina.)

How does rape affect a woman? How does rape affect those closest to a rape victim? How can those closest to a rape victim do "the right thing?" The answer to these questions is that we don't know but we have some ideas which we hope will offer a beginning for giving effective support to victims of rape. For more than anyone else, it is those closest to the victim who influence how she will deal with the attack.

What is rape? Rape is legally defined as forcible vaginal penetration to which a woman does not consent. Most women who have been raped do not react to the sexual aspects of the crime, but instead they react to the terror and fear that is involved. Often an immediate reaction of the woman is "I could have been killed." Many of those around her, particularly men, may find themselves wondering about the sexual aspects of the crime. The more this concern about sex is communicated to the woman, the more likely she is to have difficulties in dealing with her own feelings. Probably the best way to understand her feelings is to try to remember or imagine a situation where you felt powerless and afraid. You may remember feeling very alone, fearful and needing comfort. Perhaps the best way to get over these feelings is to have a comforting physical environment.

Often the raped woman needs much love and support the first few days. Affection seems to be important. Such things as holding and touching, can be really appreciated; they help relieve the loneliness. This, of course, leads to the question of sex. It is impossible to generalize about how the woman will feel about sex, nor should you guess. If you have been involved sexually with the woman, try to discuss at an appropriate time how she feels 1) about the attack, 2) about you, and 3) about sex. (An appropriate time is not right after the rape! Let her comments guide you in deciding whether you have chosen a good time to discuss it or whether you should wait until another time.) Some women will be anxious to resume normal sexual relations as a way of forgetting the rape; others will be more hesitant. However, a tendency to put off dealing with sexual relations for a long time is not typical, and appropriate help should be sought if this happens.

In the case of virgin rapes, female support seems most important. It is a good time to discuss the pleasures involved in sex—as well to reassert the woman's right to decide when and with whom she wishes to have sex. A woman's mother may feel comfortable about this; if not, a sister or friend.

It seems advisable for the woman to talk about the rape; however, it is not possible to generalize about how much she should be encouraged to talk about it. Women do not seem to appreciate specific questions about the actual rape. To probe may only worsen any problems the woman may have in dealing with the rape. Instead questions about how she feels now, and what bothers her most are more useful. They are not as threatening and will help her to talk about her most immediate concerns. Often, the rape will cause the woman to think about other problems. She may have a feeling that she has "failed" in some way. Probably the most practical suggestion is that you communicate your own willingness to let her talk about the rape. Because of your closeness to her, the woman may be more able to talk with you. However, if the rape distresses you, it may be difficult for her to talk with you. She may also want to protect you from this distress. In these and perhaps other cases, she really will not be able to talk with you— encourage her to speak with someone else she trusts. Remember the rape has brought up feelings of powerlessness—encouraging her to talk to whom she wants, when she wants to, is more helpful than making her feel that it is necessary to talk to you.

If rape is treated as a serious crime and not a sinful experience, women would probably have less difficulties in dealing with it. The woman survived the attack, and one would suppose that she would want to resume living her usual life as quickly as possible. Most women will find that memories of the rape call up other unhappy times in their lives. Memories may interfere with doing pleasant usual activities. If you notice a change, talk to the woman about what is happening to her. If, after a reasonable amount of time, a woman seems unable to cope with day-to-day problems of life, professional help should be sought.

Whether or not professional counseling is sought, remember that it is not a replacement for warm, concerned, loving care. A professional counselor may help, but he or she cannot replace those close to the victim. Rape not only affects the woman, but you, as it plays upon your own fears. Try to see the fears for what they are, and don't make them more than they are. This will help you to be able to hear what the woman is saying.

Finally, it should be noted that if the woman has pressed legal charges, this involves further painful remembering and experiences. Your awareness of the legal processes and problems involved and your support can be very helpful.

APPENDIX V

PROPOSED ADDITIONAL GUIDELINES FOR HOSPITAL CARE OF CHILD VICTIMS OF SUSPECTED OR ALLEGED RAPE

Seiden, A., and Grossman, M., Appendix to P. Bart, *et. al.* Hospital Sub-committee Report, included in *Recommendations and Report of the Citizens' Advisory Council on Rape.* Cook County, Illinois, March 10, 1975. Reprinted by permission of the authors.

I. CLASSIFICATION OF CIRCUMSTANCES

A. Rape which is both statutory and forcible is alleged to have occurred.

1. Post-pubescent patients may in general be treated as adults, with certain modifications. The rape victim's advocate should talk separately, first with the victim and then with the accompanying parent or other adult (if any) in order to decide whether the adult is emotionally prepared to be supportive to the patient during examination and interrogation. A responsible adult who is too emotionally distraught to be of help to the patient should be provided with another person with whom to talk about the situation, allowing for ventilation of feelings, and making of practical decisions.

2. Pre-pubescent patients may require special counseling skills not necessarily possessed by the rape victim's advocate who is experienced in dealing with adult women on a woman-to-woman basis. The hospital offering treatment to pre-pubescent rape victims must have available on 24-hour call a person, ordinarily a woman mental health professional or skilled paraprofessional, with specific skills and experience *both* in communication with emotionally distraught children, and with the specifics of rape counseling in children. Besides this person, the hospital must also provide some-

one skilled and experienced in counselling with the parents of children who have been raped. In many cases the regular rape victim's advocate will be able to play this role.

3. The possibility of special emotional vulnerability in younger patients must be recognized. Such patients may require tranquilizing medication (such as Valium) in order to achieve sufficient relaxation to tolerate a pelvic examination with any degree of comfort; if this is needed, it must be provided sufficiently in advance of the examination to have had time to take effect.

4. The physical surroundings for pre-pubescent children, throughout examination and interviews and any prolonged period of waiting, must be "homelike" rather than "medical," so as to avoid unnecessarily frightening the child. Specifically, there must be homelike curtains at any windows, pictures on the walls, furniture comfortable for the child, and provision of at least a few toys for any prolonged waiting period. Any necessary medical equipment of a sort unfamiliar to children must be stored out of sight in a closet or behind a screen. Comfortable and appropriate clothing must be provided for the child whose own clothing is torn and/or needed for evidence. Standard medical treatment rooms of the sort resembling operating rooms *must not* be used for children except in the event of severe trauma requiring operative treatment, and under these circumstances, the surroundings must naturally be explained to the child as well as possible by the accompanying counselor. Appropriate food must be provided for any child who is hungry when brought in, or who must be kept waiting for any prolonged period of time.

5. The physician examining the child must be a pediatrician, pediatric resident, or pediatric gynecologist, skilled in communication with traumatized children, and having received specific instruction and supervision in the treatment of pediatric rape victims. In the event that severe trauma requires treatment skills not possessed by the pediatrician, another medical specialist should be called in, and in the event that this is necessary, the pediatrician first examining the child should remain present for the specialist's evaluation and treatment, to give continuing psychological support to the child. (The pediatrician's continuing presence might be unnecessary only in the event that the counselor appears to have established a better supportive relationship with the child, and the

presence of too many strangers is adjudged to be more disturbing to the child than the pediatrician's departure.)

6. The child must not be discharged without a written plan for adequate medical and psychological follow-up. This must include written instructions to parents for medical danger signs to look for (such as evidence of further bleeding, infection, possible late evidence of undetected head injury, etc.). There must also have been some consideration of the circumstances of the rape, and a plan to attempt to correct any continuing vulnerability (is the rapist known to the family? Apprehended or still at large? Are arrangements for care and supervision of the child generally adequate? and the like). The parent must have been given some guidance if indicated on how to handle further questions the child may have, allow for helpful communication rather than further trauma, and the like. If it appears that the parent will have difficulty in being helpful to the child, further counseling as indicated for parent, or child, or both, must be planned. All follow-up arrangements must be explained to the child as far as possible, consistent with the child's ability to comprehend. Both child and parent must be given the name and telephone number of the persons responsible for further medical and psychological help, in the event that it is needed; this must be in writing for the parent. The possibility of need for further counseling to help in minimizing the traumatic effect of any subsequent legal procedures must be mentioned.

B. Statutory rape only is alleged to have occurred. Parent or guardian alleges that intercourse occurred, with consent of victim, and requests exam to aid in prosecution. Patient may or may not voluntarily agree to examination and interrogation.

1. All medical and voluntary personnel need to be aware of the fact that this is the one "rape" situation in which the "victim" herself may be at risk of being charged with delinquency, and that any admissions which she makes, which enter medical records, could potentially be used against her. This committee needs further legal advice before writing final guidelines to cover this situation; there may be circumstances in which the patient herself needs legal advice before deciding about examination and interrogation.

2. In the event that the patient refuses examination and/or interrogation, while the parent or guardian insists on it, a family

crisis with high risk of further child abuse must be adjudged to be present until proven otherwise.

a. The medical risks of injury from a forced pelvic exam on a resisting patient, in the absence of medical indications for such an exam (e.g. profuse bleeding), are not warranted.

b. The patient must be provided with the opportunity to discuss the situation with a qualified counselor under confidentiality protection applicable to psychotherapy, before further steps are taken. Such further steps might include (but are not limited to) psychiatric evaluation of either child or parent, offering the child the opportunity of temporary custody for safety and further assessment of the situation rather than returning home with an angered parent, and the like. The patient who claims that she has been falsely accused by a parent must be informed of the fact that an examination not confirming her having had intercourse might be useful evidence on her side.

c. Examination under anesthesia, of an unwilling patient, presents medical risks which are not warranted except under strong medical rather than legal indications—e.g. profuse bleeding. Under these circumstances a hospital administrator must authorize the examination, and it must be done in the presence of a board certified or eligible anesthesiologist able to deal with the hazards of forced anesthesia in an unwilling patient who is likely to have a full stomach.

d. Adequate medical and psychological follow-up must be provided after any of these circumstances. The patient must not be discharged without a written plan for such follow-up, and the patient and parent must be provided with a written record of the name and telephone number of the person responsible for carrying out the follow-up plan. All specific instructions for medical, legal, or psychological follow-up must be given in writing, because of the high likelihood of forgetting or misunderstanding, in emotionally difficult circumstances.

C. Male sexual assault victims: The guidelines above listed for pre-pubescent girls apply to the treatment of pre-pubescent boys as well, particularly including a psychosocial support, follow-up planning, and prophylactic treatment for venereal disease. In the case of post-pubescent males, the frequency of reported forced assault at present is not high enough to warrant the 24-hour availability of special advocates. However, a hospital certified for offering treatment to male victims of sexual assault must have available on call a mental health professional or paraprofessional skilled and experienced in the treatment of psychological crises arising from such assault, and of course the same guidelines apply for obtaining evidence and protection from venereal disease.

II. GENERAL COMMENTS

A. Protection of minors against pregnancy and venereal disease. The same guidelines developed for adults apply to victims of all ages, for protection against venereal disease, with the added proviso that a child or foster parent or guardian cannot necessarily be expected to have adequate information about a possible penicillin allergy. Contraception poses more difficult problems. A post-pubescent minor must be treated as an adult, with the further recognition that the risk of pregnancy may pose special problems for her if she or her parents or guardian oppose abortion on religious or other grounds. The decision about contraceptive procedures must be made with great care and individualized judgment for the patient if she is at or near the age of puberty. There is a risk that large doses of steroid hormones have different and as yet unknown adverse consequences at this age. The possibility that the patient is not yet fertile cannot of course be relied upon, and a decision whether to take this risk, use steroids, or subject the patient to the further trauma of a prophylactic dilatation and curretage must rest on careful consideration of the individual case and the acceptability to the patient and her family or guardian of abortion should pregnancy ensue. All patients in the statutory rape situation, as well as any others who acknowledge they are sexually active apart from the rape situation, must be provided with medically sound and personally acceptable contraceptive advice if they do not have it, as part of the follow-up plan.

B. The City of Chicago, from public health funds, should contract with the Visiting Nurse Association of Chicago (or other equivalent group) to provide a home visit to all child rape victims, approximately six weeks after the incident, to assess: (1) whether follow-up

VD testing has been done, (2) whether there are any persistent psychological or legal problems requiring attention, (3) whether any needed changes have been made to provide the child with sufficient protection in future, and (4) whether the emergency room contact was generally helpful to the patient and family, or whether changes need to be made. (This procedure, followed in Washington, D.C., might be desirable for adult victims also, in that experience has shown that there are at least some victims, stunned at the time, who have a need to talk about it later.)

C. It appears likely, from the above outline of special needs of child rape victims, that not every hospital meeting the criteria for treatment of adult rape victims will be able to meet the special needs of children. This would also appear to be more consistent with the often-neglected emotional needs of hospital personnel as well. Many members of hospital staffs are understandably angered or disturbed by being confronted with senseless and deliberate violence towards children, especially of a sexual sort; these events touch deeply ingrained taboos and horrors in most of us. For the comfort and effectiveness of personnel as well as patients, these situations are best handled by an experienced and well-coordinated team whose confidence in knowing they can provide a genuinely helpful service will help them deal with the inevitable feelings aroused.

> Hospital Subcommittee:
>   Chair., Pauline Bart, Ph.D.
>   University of Illinois
>   Abraham Lincoln School of Medicine
>   Department of Psychiatry and
>   Circle Campus, Department of Sociology
>
> Section on Child Victims:
>   Anne M. Seiden, M.D.
>   Director of Research
>   Institute for Juvenile Research
>
>   Marlyn Grossman, Ph.D.
>   WICCA, Inc.
>   Mile Square Health Center, Inc.

INDEX